T0137450

Power Systems

More information about this series at http://www.springer.com/series/4622

Ivo Barbi · Fabiana Pöttker

Soft Commutation Isolated
DC-DC Converters

 Springer

Ivo Barbi
Federal University of Santa Catarina
Florianópolis, Santa Catarina, Brazil

Fabiana Pöttker
Department of Electronics
Federal University of Technology—Paraná
Curitiba, Paraná, Brazil

ISSN 1612-1287 ISSN 1860-4676 (electronic)
Power Systems
ISBN 978-3-030-07148-6 ISBN 978-3-319-96178-1 (eBook)
https://doi.org/10.1007/978-3-319-96178-1

This Springer imprint is published by the registered company Springer Nature Switzerland AG
The registered company address is: Gewerbestrasse 11, 6330 Cham, Switzerland

To Antônio Barbi and Adriana Barbi.
                                    —Ivo Barbi

To my beloved husband, Douglas,
for his encouragement and support.
                                    —Fabiana Pöttker

Preface

Power electronics can be defined as the applied science dedicated to the electrical energy processing by the use of power semiconductors as switches. The use of static converters in the processing and control of electrical energy matured in the twentieth century to the point of becoming a vital technology of great economical relevance to society. There is a number of benefits of power electronics when compared to previous techniques including lower cost, high efficiency, high power density, and simplicity of control.

The static power converters, according to whether the input and output are alternating current (AC) or direct current (DC), are classified into four basic types: DC-DC converters, AC-DC converters, DC-AC converters, and AC-AC converters. The DC-DC converters are designed to control the power flow from a DC power source to another. They may be unidirectional or bidirectional, isolated or not. Their power ranges from a few watts to hundreds of kilowatts, while their voltage ranges from a few volts to tens of kilovolts.

DC-DC converters are used in personal computers, mobiles, electrical vehicles, lighting, railway systems, microgrids, medical equipment, avionics, and many other applications.

To optimize the power efficiency of isolated DC-DC converters, soft commutation is mandatory, particularly in high power density applications.

The aim of the authors is to present a detailed description and a quantitative analysis of the most common unidirectional soft-commutated isolated DC-DC converters, focusing on the soft commutation process, its quantitative and mathematical analysis, and the quantification of the switching parameters.

This book evolved from the author's long-term experience at two Brazilian Universities: the Federal University of Santa Catarina (UFSC) and the Federal University of Technology—Paraná (UTFPR) for graduate and undergraduate electrical engineering students. Hence, this book provides valuable information for power electronics engineers, graduate and undergraduate students, as well as researchers at universities and research institutes.

During the planning and writing of this book, we have incurred indebtedness to many people, particularly our graduate students who for years, through successive generations, have helped us to improve our texts, our methods of work, and especially our way of teaching.

Florianópolis, Santa Catarina, Brazil   Ivo Barbi
Curitiba, Paraná, Brazil   Fabiana Pöttker
June 2018

Acknowledgements

Ivo Barbi would like to thank and acknowledge valuable support provided by José Airton Beckhäuser Filho, Leonardo Freire Pacheco, Guilherme Martins Leandro, and Ygor Pereira Marca.

Fabiana Pöttker acknowledges the support of the Federal University of Technology—Paraná, for providing sufficient time and a three-month sabbatical to work on this book.

Contents

Abbreviations

CCM	Continuous conduction mode
CVC	Capacitor voltage clamped
DC	Direct current
DCM	Discontinuous conduction mode
FB	Full bridge
HB	Half bridge
LC	Inductor capacitor
LCT	Inductor capacitor thyristor
PFM	Pulse frequency modulation
PWM	Pulse width modulation
RC	Resistor capacitor
RCT	Resistor capacitor thyristor
RL	Resistor inductor
RLC	Resistor inductor capacitor
SRC	Series resonant converter
ZCS	Zero current switching
ZVS	Zero voltage switching

Chapter 1
Basic Electric Circuits with Switches

Nomenclature

V_i	Input DC voltage
E_1	DC voltage
v_C	Capacitor voltage
i_C	Capacitor current
v_L	Inductor voltage
v_R	Resistor voltage
v_S	Switch voltage
λ	Time constant
S	Switch
D	Diode
T, T_1 and T_2	Thyristors
I_o and $i_L(\infty)$	Steady state inductor current
$v_L(\infty)$ and $v_R(\infty)$	Steady state inductor and resistor voltages, respectively
W	Energy
t_f	Energy recovery time interval
N_1 and N_2	Transformer primary and secondary windings, respectively
L_m	Transformer magnetizing inductance
L_m'	Transformer magnetizing inductance referred to the secondary winding
v_1 and v_2	Transformer primary and secondary voltages, respectively
Δt_1 and Δt_2	Time interval 1 and 2, respectively
i_1 and i_2	Currents at time interval 1 and 2, respectively
I_1 and I_2	Initial currents at time interval 1 and 2, respectively
V_1	Input DC voltage referred to the transformer secondary winding
V_{C0} and I_{L0}	Capacitor and inductor initial conditions, respectively

© Springer International Publishing AG, part of Springer Nature 2019
I. Barbi and F. Pöttker, *Soft Commutation Isolated DC-DC Converters*,
Power Systems, https://doi.org/10.1007/978-3-319-96178-1_1

1.1 Introduction

In this chapter, basic electric circuits with switches are presented and analyzed as a basic knowledge for the next chapters, in which the static converters topological states, for every time interval, are represented by first and second order equivalent electric circuits. Furthermore, in the mathematical analysis, all semiconductors are considered ideal.

1.2 RC and RL DC Circuits

In this section, RC and RL DC circuits with switches, diodes and thyristors are analyzed.

1.2.1 RC DC Circuit with a Series Thyristor

The electric circuit studied in this section is in Fig. 1.1.

At the time t = 0 s the thyristor is fired, so the voltage source V_i is connected in series with the RC circuit. As there is only one energy storage device, this is a first order circuit. Considering that the capacitor voltage initial condition is zero $(v_C(0) = 0)$, (1.1) and (1.2) may be written.

$$V_i = v_C(t) + Ri_C(t) \tag{1.1}$$

$$i_C(t) = C\,\frac{dv_C(t)}{dt} \tag{1.2}$$

Substituting (1.2) in (1.1) gives

$$V_i = v_C(t) + RC\,\frac{dv_C(t)}{dt} \tag{1.3}$$

Fig. 1.1 Series RCT circuit

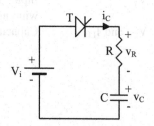

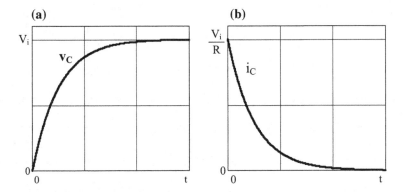

Fig. 1.2 Capacitor voltage and current waveforms

Solving (1.3), it is found

$$v_C(t) = V_i\left(1 - e^{-\frac{t}{\lambda}}\right) \tag{1.4}$$

The capacitor current is given by

$$i_C(t) = C\,\frac{dv_c(t)}{dt} = \frac{V_i}{R}\,e^{-\frac{t}{\lambda}} \tag{1.5}$$

where $\lambda = RC$ is the circuit time constant.

The capacitor voltage and current waveforms are shown in Fig. 1.2. First the capacitor is initially discharged, so the voltage V_i is applied to the resistor and the current is equal to V_i/R. As the capacitor is charged, the current decreases. When the current reaches zero in steady state (approximately at $t = 5\lambda$), the thyristor spontaneously turns off.

1.2.2   RL DC Circuit with a Series Thyristor

The electric circuit studied in this section is shown in Fig. 1.3.

At the instant $t = 0$ s the thyristor is triggered, then the voltage source V_i is connected to the series RL circuit. As the inductor is the only energy storage device, this is also a first order circuit. Considering that the inductor current initial condition is zero ($i_L(0) = 0$), (1.6) and (1.7) are obtained.

$$V_i = v_L(t) + R\,i_L(t) \tag{1.6}$$

Fig. 1.3 Series RLT circuit

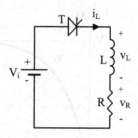

$$v_L(t) = L \frac{di_L(t)}{dt} \tag{1.7}$$

Substituting (1.7) into (1.6) gives

$$V_i = L \frac{di_L(t)}{dt} + R\, i_L(t) \tag{1.8}$$

Solving (1.8) the inductor current and voltage are found and given by (1.9) and (1.10) respectively.

$$i_L(t) = \frac{V_i}{R} \left(1 - e^{-\frac{t}{\lambda}} \right) \tag{1.9}$$

$$v_L(t) = L \frac{di_L(t)}{dt} = V_i\, e^{-\frac{t}{\lambda}} \tag{1.10}$$

where $\lambda = L/R$ is the circuit time constant.

The inductor voltage and current waveforms are shown in Fig. 1.4.

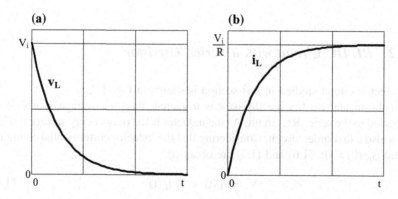

Fig. 1.4 Inductor voltage and current waveforms

1.2.3 RL DC Circuit with a Series Switch and Free Wheel Diode

The electric circuit studied in this section is illustrated in Fig. 1.5.

In the time interval 1 the switch S is closed and the diode D is reverse biased by the input voltage V_i. Current flows through switch S, the resistor and inductor, as presented in Fig. 1.6a. Therefore, Eqs. (1.11), (1.12) and (1.13) define the steady state condition ($t \gg 5\lambda$, where $\lambda = L/R$) for this time interval.

$$i_L(\infty) = I_0 = \frac{V_i}{R} \tag{1.11}$$

$$v_L(\infty) = 0 \tag{1.12}$$

$$v_R(\infty) = V_i \tag{1.13}$$

At the instant $t = 0$ s switch S is turned off. The diode D is forward biased by the inductor, initiating the time interval 2, known as the free-wheeling time interval, as shown in Fig. 1.6b. Equation (1.14) may be written for this time interval.

$$L\frac{di_L(t)}{dt} + Ri_L(t) = 0 \tag{1.14}$$

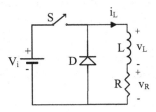

Fig. 1.5 RL DC circuit with a series switch and free wheel diode

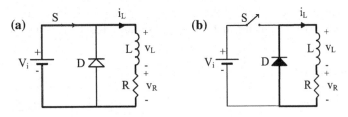

Fig. 1.6 RL DC circuit with a series switch and free wheel diode: **a** topological state for time interval 1, **b** topological state for time interval 2

Solving (1.14) gives

$$i_L(t) = I_o\, e^{-\frac{t}{\lambda}} \tag{1.15}$$

During the free-wheeling time interval, the energy accumulated in the inductor L is dissipated in the resistor R. The inductor demagnetization time depends on the value of resistance R. The higher the resistance, the faster the inductor demagnetization.

The total amount of energy dissipated in the resistor is given by

$$W = \frac{1}{2}LI_o^2 \tag{1.16}$$

If there was no free wheel diode, an unpredictable overvoltage across the switch S could be produced at the instant it is turned off.

1.3 Inductor DC Circuits

In this section, inductor DC circuits with switches, diodes and transformers are analyzed.

1.3.1 Inductor DC Circuit with a Series Switch, Free-Wheel Diode and Energy Recovery

In many applications, due to efficiency constrains, it is important to recover the energy accumulated in the inductor. A basic circuit that enables this recovery is shown in Fig. 1.7.

During the time interval 1 switch S is closed and energy is stored in the inductor, as presented in Fig. 1.8a.

At the instant t = 0 s the switch S is turned off and the inductor current is $i_L(0) = I_0$. Diode D is forward biased by the inductor, initiating the free-wheeling time interval, as shown in Fig. 1.8b. Equation (1.17) represents the time interval in which the energy accumulated in the inductor is transferred to the voltage source E_1.

Fig. 1.7 Inductor DC circuit with a series switch, a free-wheel diode and energy recovery

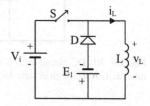

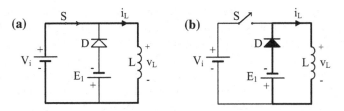

Fig. 1.8 Inductor DC circuit with a series switch, a free-wheel diode and energy recovery: **a** topological state for time interval 1, **b** topological state for time interval 2

$$L\frac{di_L(t)}{dt} = -E_1 \tag{1.17}$$

Solving (1.17) gives

$$i_L(t) = I_o - \frac{E_1}{L}t \tag{1.18}$$

At the instant $t = t_f$, given by (1.19), the inductor current reaches zero. According to this expression, the larger voltage E_1, the lower the energy recovery time interval (t_f).

$$t_f = \frac{LI_o}{E_1} \tag{1.19}$$

1.3.2 Inductor DC Circuit with a Series Switch, Free-Wheel Diode and Energy Recovery Transformer

Another circuit to recover the energy accumulated in an inductor is presented in Fig. 1.9. The transformer recovers the inductor energy to the input voltage source V_i. This technique is usually utilized in some isolated DC-DC circuits, such as the forward converter, to demagnetize the transformer.

Fig. 1.9 DC circuit with a series switch, a free-wheel diode and energy recovery transformer

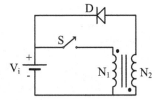

Considering an ideal transformer (no leakage inductance or parasitic resistance) the equivalent circuit is shown in Fig. 1.10, with the magnetizing inductance referred to the primary side of the transformer.

In the time interval 1, shown in Fig. 1.11, switch S is turned on and energy is transferred from the input source V_i to the transformer magnetizing inductance. Also, the diode D is reverse biased by the transformer secondary voltage.

When switch S is turned off, time interval 2 begins. At this time interval, the transformer magnetizing inductance is referred to the secondary side, as illustrated in Fig. 1.12. Then, diode D is forward biased by the transformer secondary voltage and the energy accumulated in the transformer magnetic field is delivered to the input voltage V_i.

The currents i_1 and i_2 for the time interval 1 (Δt_1) and 2 (Δt_2), respectively, are presented in Fig. 1.13, as well as the voltage across the switch S (v_S).

The initial currents I_1 and I_2 are given by

$$I_1 = \frac{V_i}{L_m} \Delta T_1 \tag{1.20}$$

$$I_2 = \frac{N_1}{N_2} I_1 \tag{1.21}$$

The expression for the current in the time interval 2 is given by

$$i_2(t) = I_2 - \frac{V_i}{L'_m} t \tag{1.22}$$

Fig. 1.10 Equivalent circuit

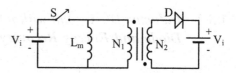

Fig. 1.11 Topological state for time interval 1

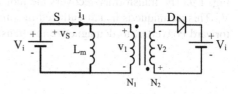

Fig. 1.12 Topological state for time interval 2

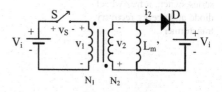

Fig. 1.13 Currents i_1 and i_2
and voltage v_S

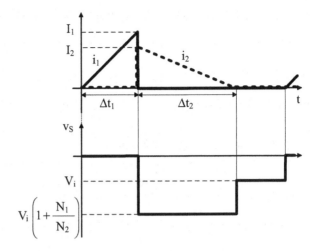

At the end of time interval 2 the current reaches zero. Thus

$$0 = I_2 - \frac{V_i}{L'_m} \Delta T_2 \tag{1.23}$$

Substituting (1.21) into (1.23), (1.24) it is obtained

$$\frac{N_1}{N_2} I_1 - \frac{V_i}{L'_m} \Delta T_2 = 0 \tag{1.24}$$

Referring the magnetizing inductance to the primary side of the transformer gives

$$\frac{N_1}{N_2} I_1 = \frac{V_i \, \Delta T_2 \, N_1^2}{L_m \, N_2^2} \tag{1.25}$$

Substituting (1.20) into (1.25), (1.26) is found

$$\frac{V_i}{L_m} L_m \Delta T_1 = V_i \, \Delta T_2 \, \frac{N_1}{N_2} \tag{1.26}$$

Hence,

$$\Delta T_2 = \frac{N_2}{N_1} \, \Delta T_1 \tag{1.27}$$

The energy recovery time ΔT_2 may be adjusted according to the transformer turns ratio, according to expression (1.27).

During the time interval 2 the voltage across switch S is given by

$$v_S = -(V_i + V_1) \tag{1.28}$$

where

$$V_1 = \frac{N_1}{N_2} V_i \tag{1.29}$$

Substituting (1.29) into (1.28) gives

$$v_S = -\left(1 + \frac{N_1}{N_2}\right) V_i \tag{1.30}$$

After the end of the interval 2 the current is zero and the voltage across switch S is $v_S = V_i$.

1.4 Constant Current Capacitor Charging Circuit

A constant current capacitor charging circuit is provided in Fig. 1.14. In the time interval 1 shown in Fig. 1.15a switch S is opened. The capacitor voltage is $v_c = 0$ and the current I flows through diode D. At the instant t = 0 s switch S is turned on then the diode D turns off and the current I starts to flow through capacitor C, as shown in Fig. 1.15b, charging it with a constant current.

During the time interval 2 the capacitor voltage is stated as

$$v_C(t) = \frac{I}{C} t \tag{1.31}$$

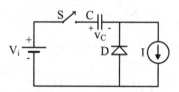

Fig. 1.14 Constant current capacitor charging circuit

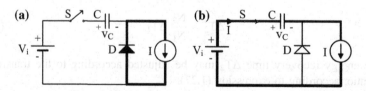

Fig. 1.15 Constant current capacitor charging circuit: **a** topological state for time interval 1, **b** topological state for time interval 2

Fig. 1.16 Voltage across
capacitor C

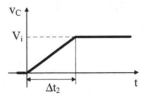

When $v_C(t) = V_i$ the diode D is forward biased and turned on. The capacitor
remains charged with the voltage V_i. The time interval 2 is calculated as follows.

$$\Delta t_2 = \frac{V_i C}{I} \tag{1.32}$$

The capacitor voltage waveform is shown in Fig. 1.16.

1.5   LC and RLC DC Circuits

In this section, LC and RLC DC circuits with switches and thyristors are analyzed.

1.5.1   LC DC Circuit with a Series Switch

The LC DC circuit with a series switch is demonstrated in Fig. 1.17. The initial
conditions are $v_C(0) = V_{Co}$ and $i_L(0) = I_{Lo}$.

At the instant $t = 0$ s switch S is turned on and the following equations may be
written:

$$V_i = v_C(t) + L\frac{di_L(t)}{dt} \tag{1.33}$$

$$i_L(t) = C\frac{dv_C(t)}{dt} \tag{1.34}$$

Fig. 1.17 LC DC circuit with
a series switch

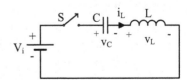

Substituting (1.34) into (1.33) gives

$$V_i = v_C(t) + LC\frac{d^2v_C(t)}{dt^2} \tag{1.35}$$

Solving (1.35), it can be obtained (1.36) and (1.37).

$$v_C(t) = -(V_i - V_{Co})\cos(\omega_o t) + I_{Lo}\sqrt{\frac{L}{C}}\operatorname{sen}(\omega_o t) + V_i \tag{1.36}$$

$$\sqrt{\frac{L}{C}}i_L(t) = (V_i - V_{Co})\operatorname{sen}(\omega_o t) + I_{Lo}\sqrt{\frac{L}{C}}\cos(\omega_o t) \tag{1.37}$$

Multiplying (1.37) by "j" and adding to (1.36) we obtain

$$v_C(t) + j\sqrt{\frac{L}{C}}i_L(t) = -(V_i - V_{Co})[\cos(\omega_o t) - j\operatorname{sen}(\omega_o t)]$$

$$+ jI_{Lo}\sqrt{\frac{L}{C}}[\cos(\omega_o t) - j\operatorname{sen}(\omega_o t)] + V_i \tag{1.38}$$

where $\omega_o = \frac{1}{\sqrt{LC}}$.

Let us define $z(t)$ and z_1 as follows:

$$z(t) = v_C(t) + j\sqrt{\frac{L}{C}}i_L(t) \tag{1.39}$$

$$z_1 = -(V_i - V_{Co}) + jI_{Lo}\sqrt{\frac{L}{C}} \tag{1.40}$$

Since

$$e^{-j\omega_o t} = \cos(\omega_o t) - j\operatorname{sen}(\omega_o t) \tag{1.41}$$

$z(t)$ can be rewrite as

$$z(t) = z_1 e^{-j\omega_o t} + V_i \tag{1.42}$$

(A) Particular Cases

(A.1) $V_{Co} = 0$, $I_{Lo} = 0$, $V_i \neq 0$

According to initial conditions of this particular case, using the previous equations we find

$$z_1 = -V_i \tag{1.43}$$

At the instant $t = 0$ s we have $z(0) = 0$.
Thus,

$$z(t) = -V_i e^{-j\omega_o t} + V_i \tag{1.44}$$

The state plane representation of this particular case is shown in Fig. 1.18.
(A.2) $I_{Lo} = 0$, $V_i = 0$, $V_{Co} > 0$
With these initial conditions and the previous equations, we have

$$z_1 = V_{Co} \tag{1.45}$$

$$z(t) = V_{Co} \tag{1.46}$$

$$z(t) = V_{Co} \, e^{-j\omega_o t} \tag{1.47}$$

The normalized state plane of this particular case is shown in Fig. 1.19.
(A.3) $V_{Co} = V_i = 0$, $I_{Lo} > 0$
With these initial conditions, Eqs. (1.48), (1.49) and (1.50) can be found.

$$z_1 = jI_{Lo}\sqrt{\frac{L}{C}} \tag{1.48}$$

$$z(0) = jI_{Lo}\sqrt{\frac{L}{C}} \tag{1.49}$$

Fig. 1.18 State plane representation for $V_{Co} = I_{Lo} = 0$ and $V_i \neq 0$

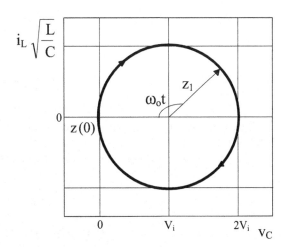

Fig. 1.19 Normalized state plane for $I_{Lo} = 0$, $V_i = 0$ and $V_{Co} > 0$

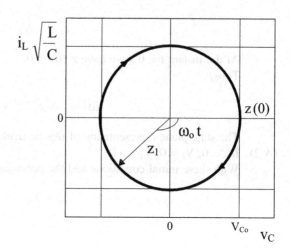

Fig. 1.20 State plane representation for $V_{Co} = V_i = 0$ and $I_{Lo} > 0$

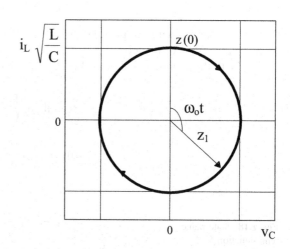

$$z(t) = jI_{Lo}\sqrt{\frac{L}{C}}e^{-j\omega_o t} \qquad (1.50)$$

The corresponding state plane representation of this particular case is shown in Fig. 1.20.

Regardless of the particular cases presented above we have

$$v_C(t) = \text{Re}\{z(t)\} \qquad (1.51)$$

$$i_L(t)\sqrt{\frac{L}{C}} = \text{Im}\{z(t)\} \qquad (1.52)$$

Substituting z(t) into (1.51) and (1.52) yields

$$v_C(t) = \text{Re}\{z_1 e^{-j\omega_0 t}\} + V_i \tag{1.53}$$

$$i_L(t)\sqrt{\frac{L}{C}} = \text{Im}\{z_1 e^{-j\omega_0 t}\} \tag{1.54}$$

1.5.2 LC DC Circuit with a Series Thyristor

The LC DC circuit with a series thyristor is illustrated in Fig. 1.21. The initial conditions are $v_C(0) = 0$ and $i_L(0) = 0$. At the instant $t = 0$ s the thyristor T is gated on. The corresponding normalized state plane trajectory of the capacitor voltage and the inductor current is shown in Fig. 1.22. The capacitor voltage and inductor current waveforms are presented in Fig. 1.23.

At the instant $t = \frac{\pi}{\omega_0}$, the inductor current reaches zero and the thyristor naturally turns off. At this instant the capacitor is charged and its voltage is equal to $2V_i$.

The following equations represent the capacitor voltage and the inductor current as function of time, respectively:

$$v_C(t) = -V_i \cos(\omega_0 t) + V_i \tag{1.55}$$

Fig. 1.21 LC DC circuit with a series thyristor

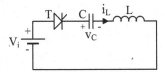

Fig. 1.22 Normalized state plane for the LCT circuit

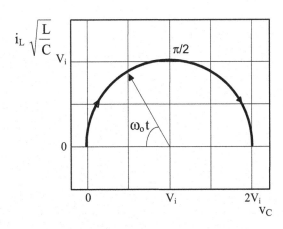

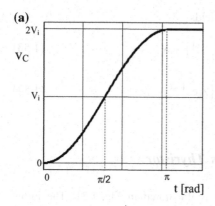 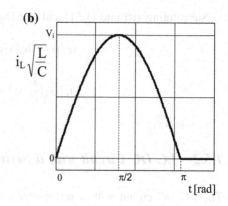

Fig. 1.23 Capacitor voltage (**a**) and inductor current (**b**) waveforms

$$i_L(t)\sqrt{\frac{L}{C}} = V_i \operatorname{sen}(\omega_0 t) \qquad (1.56)$$

1.5.3 Capacitor Voltage Polarity Inversion Circuit

A capacitor voltage polarity inversion circuit is shown in Fig. 1.24.

At t = 0 s the thyristor is triggered. The capacitor initial voltage is $v_C(0) = -V_o$. The state plane trajectory is provided in Fig. 1.25, also the capacitor voltage and inductor current waveforms are shown in Fig. 1.26. At the instant the inductor current reaches zero the capacitor voltage is $v_C\left(\frac{\pi}{\omega_0}\right) = +V_o$, with inverted polarity in relation to the initial voltage.

Fig. 1.24 Capacitor voltage
polarity inversion circuit

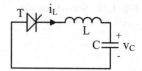

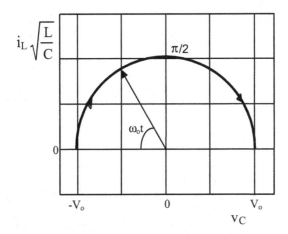

Fig. 1.25 State plane trajectory for the circuit shown in Fig. 1.24

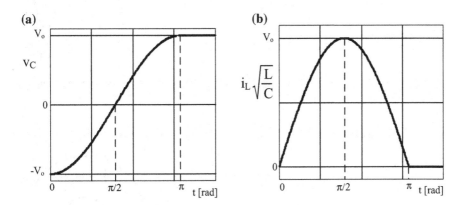

Fig. 1.26 Capacitor voltage (**a**) and inductor current (**b**) waveforms

1.5.4 Capacitor Charging Circuits

(A) First Circuit

The circuit analyzed in this section is shown in Fig. 1.27.

Thyristor T_1 and T_2 are gated on at a switching frequency lower than the resonant frequency. The capacitor voltage and the inductor current waveforms are shown in Fig. 1.28 and the state plane trajectory is presented in Fig. 1.29. To simplify, all components are considered ideal. Consequently, the circuit damping factor is zero and the energy transferred to the capacitor rises over time indefinitely (at least theoretically).

Fig. 1.27 Capacitor charging circuit

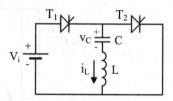

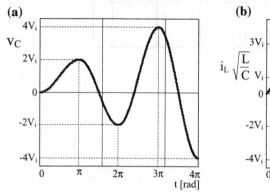

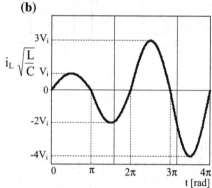

Fig. 1.28 Capacitor voltage (**a**) and parameterized inductor current (**b**) waveforms over time

Fig. 1.29 State plane trajectory for the circuit shown in Fig. 1.27

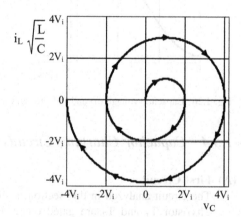

(B) Second Circuit

Now, analyzing the operation of the circuit shown in Fig. 1.30, with initial conditions $V_{Co} < 0$ and $I_{Lo} = 0$. At $t = 0$ s, T_1 is gated on. The capacitor voltage and current evolves over time and before the current reaches zero T_2 turned on, switching off T_1. Therefore, energy is transferred from the input

Fig. 1.30 Capacitor charger
circuit

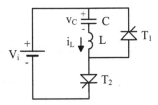

voltage source V_i to the capacitor C, rising its voltage above V_i. The capacitor
voltage and inductor current waveforms are shown in Fig. 1.31.

During the time interval T_1 is closed, the capacitor voltage is given by

$$v_C(t) = -V_{Co} \cos(\omega_o t) \tag{1.57}$$

At the end of this interval the capacitor voltage and current are given by
Eqs. (1.58) and (1.59) respectively.

$$V_1 = -V_{Co} \cos(\omega_o \tau) \tag{1.58}$$

$$\sqrt{\frac{L}{C}} I_1 = V_{Co} \operatorname{sen}(\omega_o \tau) \tag{1.59}$$

When T_2 is turned on we have

$$z_1 = (V_1 - V_i) + jI_1 \sqrt{\frac{L}{C}} \tag{1.60}$$

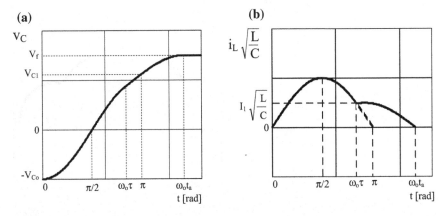

Fig. 1.31 Capacitor voltage (**a**) and inductor current (**b**) waveforms for the circuit shown in
Fig. 1.30

$$z(0) = V_1 + jI_1 \sqrt{\frac{L}{C}} \qquad (1.61)$$

At the end of this time interval the capacitor voltage is given by

$$V_f = V_i + |z_1| \qquad (1.62)$$

Substituting (1.58) and (1.59) into (1.60) yields

$$|z_1|^2 = (V_{Co} \cos(\omega_o \tau) - V_i)^2 + V_{Co}^2 \operatorname{sen}^2(\omega_o \tau) \qquad (1.63)$$

Substituting (1.63) into (1.62) gives

$$V_f = V_i + \sqrt{(V_{Co} \cos(\omega_o \tau) - V_i)^2 + V_{Co}^2 \operatorname{sen}^2(\omega_o \tau)} \qquad (1.64)$$

According to Eq. (1.64) the capacitor voltage is controlled by the angle $\omega_o \tau$. In the particular case which $\omega_o \tau = \pi$, the capacitor final voltage value (Vf) is given by

$$V_f = V_i + \sqrt{(-V_{Co} - V_i)^2} = V_i - V_i - V_{Co} \qquad (1.65)$$

or,

$$V_f = -V_{Co} \qquad (1.66)$$

This circuit is used in auxiliary commutation circuits as well as in resonant converters. The state plane trajectory is shown in Fig. 1.32.

Fig. 1.32 State plane trajectory for the circuit shown in Fig. 1.31

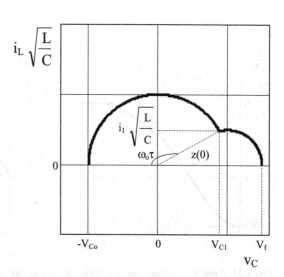

1.5.5   RLC Circuit with a Low Dumping Factor

Let us consider the circuit shown in Fig. 1.33. The capacitor current and voltage are given by (1.67) and (1.68) respectively.

$$i_C(t) = \frac{V_i - V_o}{\omega L} e^{-\alpha t} \text{sen}(\omega t) - I_o \frac{\omega_o}{\omega} e^{-\alpha t} \text{sen}(\omega t - \gamma) \qquad (1.67)$$

$$v_C(t) = V_i - (V_i - V_o) \frac{\omega}{\omega_o} e^{-\alpha t} \text{sen}(\omega t + \gamma) + \frac{I_o}{\omega C} e^{-\alpha t} \text{sen}(\omega t) \qquad (1.68)$$

where

$$\omega_o = \frac{1}{\sqrt{LC}} \qquad (1.69)$$

$$\alpha = \frac{R}{2L} \qquad (1.70)$$

$$\gamma = \text{arc tg}\left(\frac{\omega}{\alpha}\right) \qquad (1.71)$$

$$\omega^2 = \omega_o^2 - \alpha^2 \qquad (1.72)$$

Assuming that the circuit quality factor is high, what is true for very small losses, we have

$$\omega_o \cong \omega \qquad (1.73)$$

Let

$$X = \sqrt{\frac{L}{C}} = \omega L \cong \frac{1}{\omega C} \qquad (1.74)$$

$$\psi = \frac{X}{R} \qquad (1.75)$$

$$\frac{\alpha}{\omega} = \frac{R}{2\omega L} = \frac{1}{2\psi} \qquad (1.76)$$

Fig. 1.33 RLC circuit with a
low dumping factor

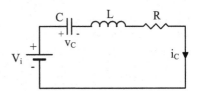

$$\gamma = \frac{\pi}{2} \tag{1.77}$$

$$\text{sen}(\omega t - \gamma) = -\cos(\omega t) \tag{1.78}$$

Manipulating Eqs. (1.74), (1.75), (1.76), (1.77) and (1.78) gives

$$i_L(t) = \left(\frac{V_i - V_o}{X}\,\text{sen}\,(\omega t) + I_o \cos\,(\omega t)\right) e^{-\frac{\omega t}{2\psi}} \tag{1.79}$$

and

$$v_C(t) = V_i + [X I_o \,\text{sen}\,(\omega t) - (V_i - V_o)\cos\,(\omega t)]\, e^{-\frac{\omega t}{2\psi}} \tag{1.80}$$

where $e^{-\frac{\omega t}{2\psi}} = e^{-\alpha t}$ can be represented by the series

$$e^{-\alpha t} = 1 - \alpha t + \frac{\alpha^2 t^2}{2} - \frac{\alpha^3 t^3}{6} \tag{1.81}$$

For small values of α the following statement is true

$$e^{-\alpha t} = 1 - \alpha t \tag{1.82}$$

We know that

$$z(t) = v_C(t) + j\sqrt{\frac{L}{C}} i_L(t) \tag{1.83}$$

Consequently, with appropriate algebraic manipulation it is found

$$z(t) = V_i + z_1\, e^{-j\omega t}\, e^{-\alpha t} \tag{1.84}$$

In an ideal circuit $\alpha = 0$, so the Eq. (1.84) becomes equal to Eq. (1.42).

1.5.6   LC Circuit with Voltage and Current Source

A LC circuit with voltage and current sources is shown in Fig. 1.34.

Fig. 1.34 LC circuit with voltage and current sources

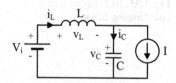

Applying Kirchhoff's voltage and current law in the circuit, we obtain

$$V_i = v_L(t) + v_C(t) \tag{1.85}$$

and

$$i_L(t) = i_C(t) + I \tag{1.86}$$

The instantaneous inductor voltage and capacitor current are given by Eqs. (1.87) and (1.88) respectively.

$$v_L(t) = L\frac{di_L(t)}{dt} = L\frac{d(I + i_C(t))}{dt} = L\frac{di_C(t)}{dt} \tag{1.87}$$

$$i_C(t) = C\frac{dv_C(t)}{dt} \tag{1.88}$$

Substituting (1.88) into (1.87) yields

$$v_L(t) = LC\frac{d^2v_C(t)}{dt^2} \tag{1.89}$$

Replacing (1.89) into (1.85) leads to

$$V_i = LC\frac{d^2v_C(t)}{dt^2} + v_C(t) \tag{1.90}$$

Rearranging Eq. (1.90) gives

$$\frac{d^2v_C(t)}{dt^2} + \frac{v_C(t)}{LC} = \frac{V_i}{LC} \tag{1.91}$$

The solution of the differential Eq. (1.91) yields

$$v_C(t) = (V_{Co} - V_i)\cos(\omega_o t) + \sqrt{\frac{L}{C}}(I_{Lo} - I)\operatorname{sen}(\omega_o t) + V_i \tag{1.92}$$

Substitution of Eq. (1.92) into (1.88) and (1.87), with appropriate algebraic manipulations gives

$$\sqrt{\frac{L}{C}}i_L(t) = -(V_{Co} - V_i)\,\text{sen}\,(\omega_o t) + \sqrt{\frac{L}{C}}(I_{Lo} - I)\cos\,(\omega_o t) + \sqrt{\frac{L}{C}}I \quad (1.93)$$

The value of $z(t)$ is given by

$$z(t) = v_C(t) + j\sqrt{\frac{L}{C}}i_L(t) \quad (1.94)$$

Replacing $\sqrt{\frac{L}{C}}\,i_L(t)$ and $v_C(t)$, that were obtained in Eqs. (1.92) and (1.93), into Eq. (1.94), gives

$$z(t) = \left(V_i + j\sqrt{\frac{L}{C}}I\right) + \left[(V_{Co} - V_i) + j\sqrt{\frac{L}{C}}(I_{Lo} - I)\right]e^{-j\omega_o t} \quad (1.95)$$

The quantities of z_o and z_1 are given by Eqs. (1.96) and (1.97):

$$z_o = V_i + j\sqrt{\frac{L}{C}}I \quad (1.96)$$

$$z_1 = (V_{Co} - V_i) + j\sqrt{\frac{L}{C}}(I_{Lo} - I) \quad (1.97)$$

Substitution of (1.96) and (1.97) into (1.95) gives

$$z(t) = z_o + z_1\,e^{-j w_o t} \quad (1.98)$$

The state plane trajectory is illustrated in Fig. 1.35.

Fig. 1.35 State plane trajectory for the circuit shown in Fig. 1.34

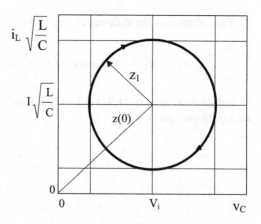

Fig. 1.36 Parallel LC circuit connected to a DC current source

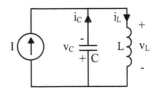

(A) First Particular Case: I = 0

Assuming that I = 0, it can be obtained Eq. (1.99) from Eq. (1.97).

$$z(t) = V_i + \left[(V_{Co} - V_i) + j\sqrt{\frac{L}{C}} I_{Lo} \right] e^{-j\omega_o t} \qquad (1.99)$$

This particular case was analyzed in Sect. 1.2.1 of this Chapter.
Second Particular Case: $V_i = 0$
Making $V_i = 0$ and substituting in Eq. (1.97) yields

$$z(t) = j\sqrt{\frac{L}{C}} I + \left[(V_{Co} - V_i) + j\sqrt{\frac{L}{C}} (I_{Lo} - I) \right] e^{-j wt} \qquad (1.100)$$

Equation (1.100) represents the behavior of a circuit constituted by the parallel association of a capacitor C, an inductor L and a current source I, illustrated in Fig. 1.36.

1.6 Solved Problem

The circuit is presented in Fig. 1.37. Assume that the thyristor T is gated on at the instant t = 0 s. The parameters of the circuit are L = 30 μH, C = 120 μF and $V_{Co}(0) = -75$ V. Find the equations for i(t), $v_L(t)$, $v_C(t)$ and $i_D(t)$ and plot them as a function of time.

Fig. 1.37 Exercise 3

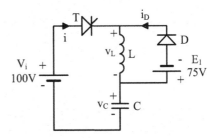

Fig. 1.38 First time interval
equivalent circuit

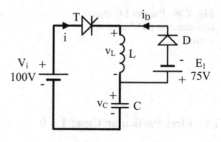

First Time Interval ($0 \leq t \leq t_1$): The equivalent circuit for this interval is shown in Fig. 1.38. It is a second order LC circuit. Substituting the values of C, L and $V_{Co}(0)$ in Eqs. (1.36) and (1.37) we find

$$v_C(t) = V_i + (V_{Co} - V_i)\cos(\omega_o t) = 100 - 175 \cos(\omega_o t)$$

$$\sqrt{\frac{L}{C}}\, i(t) = -(V_{Co} - V_i)\sin(\omega_o t) = 175 \sin(\omega_o t)$$

$$v_L(t) = -v_C(t) + V_i = 175 \cos(\omega_o t)$$

According to the inductor voltage expression, this time interval ends at the instant $t = 120.8$ µs, when $v_L(t) = -75$ V and diode D is turned on. At this instant $v_C(t) = 175$ V.

Second Time Interval ($t_1 \leq t \leq t_0$): The second time interval begins at the instant diode D is forward biased. The inductor current decays linearly and delivers energy to E_1 until the instant it reaches zero. The second time interval equivalent circuit is shown in Fig. 1.39.

The results obtained by simulation are shown in Fig. 1.40 as a function of time, while the corresponding state plane trajectory is represented in Fig. 1.41.

Fig. 1.39 Second time
interval equivalent circuit

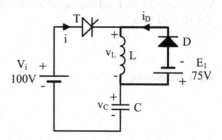

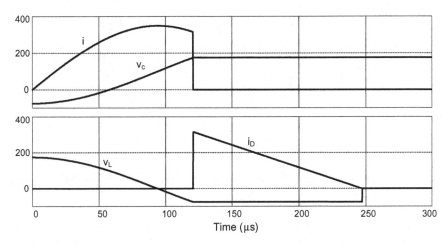

Fig. 1.40 Simulation results

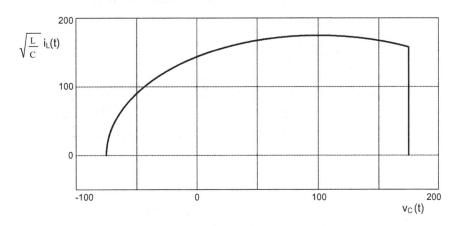

Fig. 1.41 State plane trajectory

1.7 Problems

1. Describe the operation of the circuits shown in Fig. 1.42, considering $L = 100$ μH and $C = 25$ μF, and plot the waveforms of $i(t)$, $v_{L(t)}$ and $v_{C(t)}$. The initial voltages of the capacitor are: (a) $v_C(0) = 0$ V, (b) $v_C = -50$ V, and (c) $v_C = -50$ V. The initial current in the inductor is zero. Verify the results by simulation.

   Answers: (a) $z(t) = -100e^{-j\omega_o t} + 100$;   (b) $z(t) = -150e^{-j\omega_o t} + 100$;
   (c) $z(t) = -150e^{-j\omega_o t} + 100$

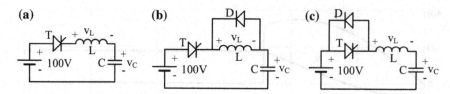

Fig. 1.42 Exercise 1

2. Describe the operation of the circuits shown in Fig. 1.43, considering L = 100 μH, C = 25 μF and $v_C(0) = -100$ V.
   Answers: (a) $z(t) = -100e^{-j\omega_o t}$; (b) $z(t) = -100e^{-j\omega_o t}$; (c) $z(t) = -100e^{-j\omega_o t}$; (d) $z(t) = -100e^{-j\omega_o t}$.
3. The capacitor initial voltage of the circuit shown in Fig. 1.44 is zero. Thyristor T_1 and T_2 are gated on periodically at a switching frequency lower than the resonant frequency of the pair LC, so they turn off spontaneously every half

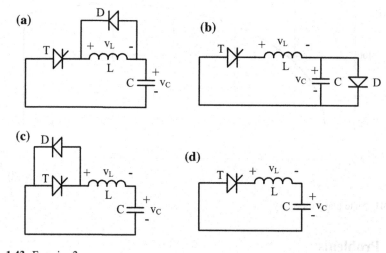

Fig. 1.43 Exercise 2

Fig. 1.44 Exercise 3

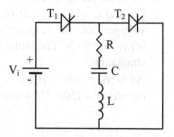

period of the switching period. The parameters of the circuit are $L = 200 \ \mu H$, $C = 20 \ \mu F$ and the quality factor is $Q = 5$. Determine: (a) the capacitor final voltage after the circuit reaches steady-state (theoretically after an infinite number of operation cycles); (b) find $v_{C(t)}$ and $i_{L(t)}$, and plot them against the time and show the state plane trajectory.

Answers: $v_{Cmin} = -270$ V and $v_{Cmax} = 370$ V.

4. The thyristor T of the circuit shown in Fig. 1.45 is turned on at the instant $t = 0$ s. The parameters of the circuit are $C = 300 \ \mu F$ with $V_{Co} = 0$ and $I_{Lo} = 0$. Find the inductance of the inductor L that results in a derivative of current $di_L/dt = 100$ A/μs.

Answer: $L = 6 \ \mu H$.

5. The number of turns in the primary and secondary windings of the transformer shown in Fig. 1.46 are $N_1 = 100$ and $N_2 = 200$ respectively. The switch S, initially closed for a long time interval, is opened at the instant $t = 0$ s. The transformer magnetizing inductance is 200 μH. Obtain the equation for the current in the secondary winding and the voltage across the switch S. Plot the relevant waveforms as a function of time and use simulation to verify the calculations.

6. Consider the circuit shown in Fig. 1.47. Thyristors T_1 and T_2 are gated on complementarily to T_3 and T_4 respectively. The switching frequency is lower

Fig. 1.45 Exercise 4

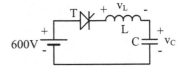

Fig. 1.46 Exercise 5

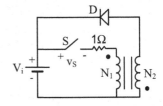

Fig. 1.47 Exercise 6

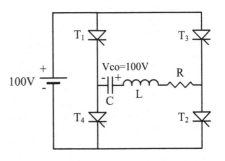

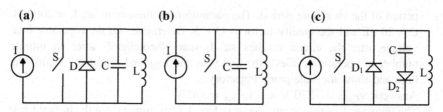

Fig. 1.48 Exercise 7

Fig. 1.49 Exercise 8

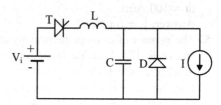

than the resonant frequency and the thyristors turn off spontaneously. The parameters of the circuit are $V_{Co} = -100$ V, $V_i = 100$ V, $I_{Lo} = 0$ A and $\alpha = 10°$. Determine the voltage across the capacitor C after the circuit reaches steady-state operation. The operation of the circuit initiates at the instant that T_1 and T_2 are gated on.

Answer: $V_c = 4$ kV.

7. Consider the circuits shown in Fig. 1.48. The switch S is initially closed for a long time and then is opened at the instant $t = 0$ s. Describe the operation of the circuits and plot the respective state planes.

8. Let the circuit shown in Fig. 1.49. At the instant $t = 0$ s the thyristor is turned on. Analyze the circuit and obtain the capacitor voltage (v_C) and the inductor current (i_L) waveforms as well as the state plane trajectory. Consider that the initial conditions are zero and $I\sqrt{L/C} < V_i$.

9. The circuits shown in Fig. 1.50a, b are initially with the switch S closed and conducting the current I. At the instant $t = 0$ s the switch S is opened. Describe

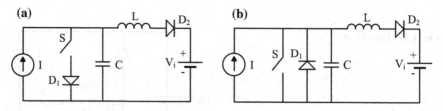

Fig. 1.50 Exercise 9

Fig. 1.51 Exercise 10

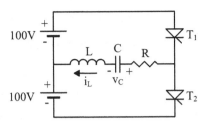

Fig. 1.52 Exercise 11

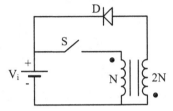

the circuits operation and obtain equations for $v_C(t)$ and $i_L(t)$. Plot the relevant waveforms against the time, along with the corresponding state planes. The initial conditions are $V_{co} = 0$ and $I_{Lo} = 0$.

10. In the circuit shown in Fig. 1.51, T_1 and T_2 are gated on complementarily with a switching frequency of 6 kHz. The parameters of the circuit are $L = 100\ \mu H$, $C = 5\ \mu F$ and $R = 0.447\ \Omega$. Consider the circuit operating in steady-state.

(a) Describe the operation of the circuit, representing the different topological states;
(b) Find the equations for the inductor current $i_L(t)$ and the capacitor voltage $v_C(t)$ and plot their waveforms against the time;
(c) Determine the inductor peak current and capacitor peak voltage.
(d) Calculate the power dissipated in resistor R.

Answers:

(b) $\qquad V_{Cmax} = \dfrac{V_i(1 + 2e^{-\alpha\pi/\omega} + e^{-2\alpha\pi/\omega})}{(1 - e^{-2\alpha\pi/\omega})};$ $\qquad i_{Lmax} = \dfrac{(V_i + V_{Cmin})}{\omega L} e^{-\alpha\pi/2\omega_0}$

(c) $V_{Cmax} = 1.27\,kV$; $i_{Lmax} = 283\,A$; (d) $P_R = 15.15\,kW$.

11. The parameters of the circuit shown in Fig. 1.52 are $V_i = 100\ V$ and $L = 1\ H$, where L is the transformer magnetizing inductance reflected to the primary winding. Switch S is closed for a time interval $t_1 = 1\ s$ and then opened. Find the time necessary time (t_2) to completely demagnetize the transformer. Answer: $t_2 = 0.25\ s$.

Chapter 2
Series Resonant Converter

Nomenclature

V_i	Input DC voltage
V_1	Half of the DC input voltage
V_o	Output DC voltage
P_o	Nominal output power
$P_{o\ min}$	Minimum output power
C_o	Output filter capacitor
R_o	Output load resistor
$R_{o\ min}$	Minimum output load resistor
q, q_{co}, q_{c1}	Static gain
D	Duty cycle
f_s	Switching frequency (Hz)
ω_s	Switching frequency (rad/s)
$f_{s\ min}$	Minimum switching frequency (Hz)
$f_{s\ max}$	Maximum switching frequency (Hz)
T_s	Switching period
f_o	Resonant frequency (Hz)
ω_o	Resonant frequency (rad/s)
μ_o	Frequency ratio (f_s/f_o)
ρ	Frequency ratio
t_d	Dead time
T	Transformer
n	Transformer turns ratio
N_1 and N_2	Transformer windings
V_o'	Output DC voltage referred to the transformer primary side
i_o	Output current
I_o'	Output current referred to the transformer primary side
$I_o'\ \left(\overline{I_o'}\right)$	Average output current referred to the transformer primary side and its normalized value in CCM

© Springer International Publishing AG, part of Springer Nature 2019

I. Barbi and F. Pöttker, *Soft Commutation Isolated DC-DC Converters*,

Power Systems, https://doi.org/10.1007/978-3-319-96178-1_2

$\left(\overline{I'_{o\,max}}\right)$	Maximum normalized average output current referred to the transformer primary in CCM
$\left(\overline{I'_{o\,min}}\right)$	Minimum normalized average output current referred to the transformer primary in CCM
$I'_{oD}\left(\overline{I'_{oD}}\right)$	Average output current in DCM referred to the transformer primary side and its normalized value
I'_{oSC}	Short circuit average output current
S_1, S_2, S_3 and S_4	Switches
D_1, D_2, D_3 and D_4	Diodes
v_{g1} and v_{g2}	Switches gate signals
L_r	Resonant inductor (may include the transformer leakage inductance)
C_r	Resonant capacitor
$v_{Cr}\left(\overline{v_{Cr}}\right)$	Resonant capacitor voltage
V_{Co}	Resonant capacitor peak voltage
V_{C1}	Resonant capacitor voltage at the end of time interval 1 and three
$i_{Lr}\left(\overline{i_{Lr}}\right)$	Resonant inductor current and its normalized value
$I_{Lr}\left(\overline{I_{Lr}}\right)$	Inductor peak current and its normalized value, commutation
I_L	Fundamental inductor peak current
$I_1\left(\overline{I_1}\right)$	Inductor current at the end of the first and third step of operation and its normalized value in CCM
I_{p1} and I_{p2} $\left(\overline{I_{p1}} \text{ and } \overline{I_{p2}}\right)$	Inductor peak current and its normalized value in DCM
v_{ab}	Full bridge ac voltage, between points "a" and "b"
v_{cb}	Inductor voltage, between points "c" and "b"
v_{ab1}	Fundamental ac voltage, between points "a" and "b"
v_{cb1}	Fundamental inductor voltage, between points "c" and "b"
v_{ac}	Voltage at the ac side of the rectifier, between points "a" and "c"
v_{S1}, v_{S2}	Voltage across switches
i_{S1}, i_{S2}	Current in the switches
Δt_1	Time interval one (t_1-t_0)
Δt_2	Time interval two (t_2-t_1)
Δt_3	Time interval three (t_3-t_2)
Δt_4	Time interval four (t_4-t_3)
Δt_5	Time interval five (t_5-t_4)
Δt_6	Time interval six (t_6-t_5)
A_1 and A_2	Area
x_{Lr}, x_{CR} and x	Reactance

Q	Capacitor charge
z	Characteristic impedance
R_1, R_2	State plane radius
ϕ_r, ϕ_o, β, θ	State plane angles

2.1 Introduction

The series resonant converter was proposed by Francisc C. Schwarz in 1975 [1] for thyristor based DC–DC converters. Due to the sinusoidal current waveforms, the thyristor naturally turns off at the instant that current reaches zero, without the need of auxiliary commutation circuits, which usually includes auxiliary thyristors, inductors and capacitors.

With the improvement of power semiconductor switches, first the Bipolar [2] and then MOSFET's and IGBT's, the series resonant converter became a very attractive option for switched mode power supply not only because it became possible to operate at high switching frequency increasing the power density, but also due to reduced commutation losses.

The series resonant converters are employed in many practical applications, which include:

- High voltage and low current power supplies for radar, X-ray and laser applications;
- Electric vehicles on-board battery chargers;
- Converters for wireless power transfer;
- DC–DC converters for traction applications;
- Solid-State Transformers (SST).

Besides its historical and technical relevance, the series resonant converter gave rise to a large family of high density resonant mode DC–DC converters, commonly used in many practical applications.

In this chapter, the series resonant converter is studied. The chapter is organized to present at first the converter topology, followed by operation description and detailed analysis in continuous and discontinuous conduction modes.

The schematics of the half-bridge and full-bridge series resonant converters are shown in Figs. 2.1 and 2.2, respectively. It consists of a full-bridge or half-bridge inverter, a resonant inductor (L_r), a resonant capacitor (C_r), a transformer, a full bridge diode rectifier and a capacitive output filter.

The series-resonant converter (SRC) is a PFM (pulse-frequency modulated) converter, since it uses a variable frequency to control the power transfer from the input to the output. In DCM (discontinuous conduction mode), ZCS (zero current switching) is achieved for both commutations of each switch, whereas in CCM (continuous conduction mode) zero current switching is achieved only when the switches turn off.

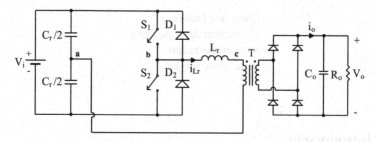

Fig. 2.1 Half-bridge series resonant converter

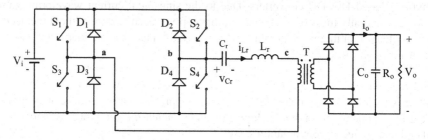

Fig. 2.2 Full-bridge series resonant converter

2.2 Circuit Operation in Continuous Conduction Mode

In this section, the half-bridge series resonant converter shown in Fig. 2.3, that is equivalent to the topology represented in Fig. 2.1, is analyzed. For the sake of simplification, we assume that [3]:

- all components are ideal and the transformer magnetizing inductance is ignored;
- the converter is in steady-state operation;
- the output filter is represented by a DC voltage source V_o', whose value is the output voltage referred to the primary side of the transformer;

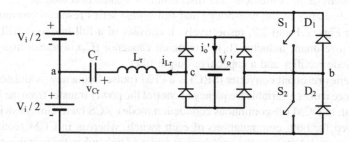

Fig. 2.3 Half-bridge series resonant converter

- current flows through the switches is unidirectional, i.e. the switches allow the current to flow only in the direction of the arrow when indicated;
- the converter is controlled by frequency modulation and the switches S_1 and S_2 are triggered with 50% duty cycle without dead-time.

In continuous conduction mode (CCM), the switches turn on is dissipative and turn off is non-dissipative (ZCS). As it will be demonstrated in the next sections, the converter operates in CCM when $0.5 f_o \leq f_s \leq f_o$, where f_o and f_s are the resonant and the switching frequencies, respectively.

The resonant capacitor peak voltage depends on the switching frequency, and it may be larger than the DC bus voltage.

(A) Time Interval Δt_1 ($t_0 < t < t_1$)

Before this time interval, switch S_2 is conducting the resonant inductor current. At $t = t_0$ the inductor current reaches zero and the diode D_2 starts conducting, as illustrated in Fig. 2.4. During this time interval energy is delivered to both the DC bus and the load. Switch S_2 must be turned off during this time interval to achieve soft commutation. This time interval finishes at the instant $t = t_1$, when switch S_1 is gated on.

(B) Time Interval Δt_2 ($t_1 < t < t_2$)

This time interval, shown in Fig. 2.5, stars at $t = t_1$ when S_1 is gated on. Forced commutation of the current takes place, from the diode D_2 to the switch S_1. The resonant capacitor C_r is discharged and charged again with opposite polarity.

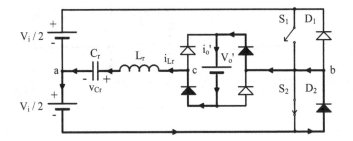

Fig. 2.4 Topological state during time interval Δt_1

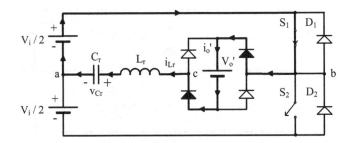

Fig. 2.5 Topological state during time interval Δt_2

The inductor current evolves with a sinusoidal waveform until it reaches zero at the instant t = t₂, when this time interval finishes. At this instant the resonant capacitor voltage is V_{C0}. During this interval the DC bus delivers energy to the load.

(C) Time Interval Δt₃ (t₂ < t < t₃)

This time interval begins at t = t₂, when the inductor current reaches zero and the diode D_1 starts to conduct, as provided in Fig. 2.6. During this time interval energy is delivered to both the DC bus and the load. Switch S_1 must be turned off during this time interval to achieve soft commutation.

(D) Time Interval Δt₄ (t₃ < t < t₄)

Figure 2.7 shows the interval Δt₄. It begins at t = t₃ when switch S_2 is gated on. Once again, forced commutation of the current takes place, from diode D_1 to switch S_2. Hence, this commutation is dissipative. The resonant capacitor C_r is discharged and charged again with opposite polarity. The inductor current evolves with a sinusoidal waveform until it reaches zero at the end of this time interval. At this instant the resonant capacitor voltage is $-V_{C0}$. During this time interval, the DC bus delivers energy to the load. The main waveforms along with time diagram are shown in Fig. 2.8.

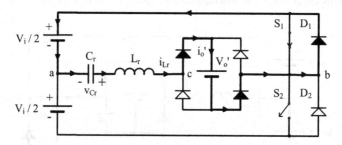

Fig. 2.6 Topological state during time interval Δt₃

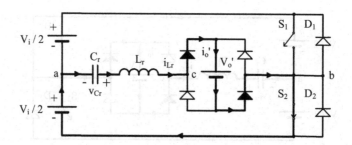

Fig. 2.7 Topological state for time interval Δt₄

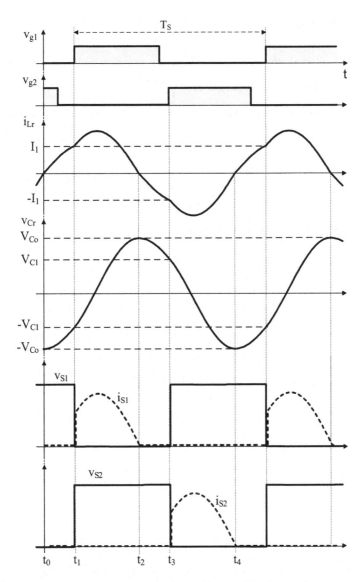

Fig. 2.8 Main waveforms and time diagram in continuous conduction mode

2.3 Mathematical Analysis for Operation in Continuous Conduction Mode

In this section, the resonant capacitor voltage and resonant inductor current expressions are obtained. Since the converter is symmetrical, only half switching period is analyzed.

2.3.1   Time Interval Δt_1

In this time interval, the initial current in the resonant inductor and the initial voltage across the resonant capacitor are: $\begin{cases} i_{Lr}(t_0) = 0 \\ v_{Cr}(t_0) = -V_{C0} \end{cases}$

From the equivalent circuit shown in Fig. 2.4, we find Eqs. (2.1) and (2.2).

$$\frac{V_i}{2} = -L_r \frac{di_{Lr}(t)}{dt} - v_{Cr}(t) - V_o' \tag{2.1}$$

$$i_{Lr}(t) = C_r \frac{dv_{Cr}(t)}{dt} \tag{2.2}$$

Applying the Laplace Transform yields

$$\frac{(V_i/2) + V_o'}{s} = -s\,L_r\,i_{Lr}(s) - v_{Cr}(s) \tag{2.3}$$

$$i_{Lr}(s) = s\,C_r\,v_{Cr}(s) + C_r\,V_{C0} \tag{2.4}$$

Substituting (2.4) in (2.3) and applying the inverse Laplace Transform we obtain,

$$v_{Cr}(t) = -\left(-V_1 - V_o' + V_{C0}\right)\cos(\omega_o t) - V_1 - V_o' \tag{2.5}$$

and

$$i_{Lr}(t)z = \left(-V_1 - V_o' + V_{C0}\right)\mathrm{sen}(\omega_o t) \tag{2.6}$$

where:

$V_1 = \frac{V_i}{2}$, $\omega_o = \frac{1}{\sqrt{L_r C_r}}$ and $z = \sqrt{\frac{L_r}{C_r}}$.

Defining the variable $z_1(t)$ as follows:

$$z_1(t) = v_{Cr}(t) + j\sqrt{\frac{L_r}{C_r}}\,i_{Lr}(t) \tag{2.7}$$

Substitution of Eqs. (2.5) and (2.6) into Eq. (2.7), with appropriate algebraic manipulation, results in

$$z_1(t) = -V_1 - V_o' + \left(V_1 + V_o' - V_{C0}\right)e^{-j\omega_o t} \tag{2.8}$$

Fig. 2.9 State-plane for the time interval Δt_1

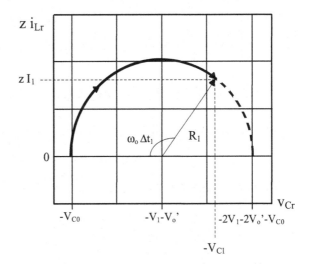

Equation (2.8) allows us to plot the state-plane for time interval Δt_1, shown in Fig. 2.9, which the coordinates of the center of the circle are $(0, -V_1 - V_o')$ and the radius is given by (2.9).

$$R_1 = V_{C0} - V_1 - V_o' \tag{2.9}$$

2.3.2 Time Interval Δt_2

In this time interval, the initial current in the resonant inductor and the initial voltage across the resonant capacitor are: $\begin{cases} i_{Lr}(t_0) = I_1 \\ v_{Cr}(t_0) = -V_{C1} \end{cases}$

The equivalent circuit presented in Fig. 2.5 yields the following equations:

$$V_1 = L_r \frac{di_{Lr}(t)}{dt} + v_{Cr}(t) + V_o' \tag{2.10}$$

$$i_{Lr}(t) = C_r \frac{dv_{Cr}(t)}{dt} \tag{2.11}$$

Applying the Laplace Transforming in Eqs. (2.10) and (2.11) we obtain the following equations:

$$\frac{V_1 - V_o'}{s} = s L_r i_{Lr}(s) - L_r I_1 + v_{Cr}(s) \tag{2.12}$$

$$i_{Lr}(s) = s C_r v_{Cr}(s) + C_r V_{C1} \tag{2.13}$$

Substituting (2.13) in (2.12) and applying the inverse Laplace Transform yields Eqs. (2.14) and (2.15).

$$v_{Cr}(t) = -\left(V_1 - V_o' + V_{C1}\right)\cos\left(\omega_o t\right) + I_1\, z \operatorname{sen}\left(\omega_o t\right) + V_1 - V_o' \tag{2.14}$$

$$i_{Lr}(t)\, z = \left(V_1 - V_o' + V_{C1}\right)\operatorname{sen}\left(\omega_o t\right) + I_1\, z \cos\left(\omega_o t\right) \tag{2.15}$$

Defining $z_2(t)$ by Eq. (2.16):

$$z_2(t) = v_{Cr}(t) + j\, z\, i_{Lr}(t) \tag{2.16}$$

Substituting Eqs. (2.14) and (2.15) in Eq. (2.16) and rearranging terms, we have

$$z_2(t) = V_1 - V_o' - \left(V_1 - V_o' + V_{C1} + j I_1\, z\right)e^{-j\omega_o t} \tag{2.17}$$

Equation (2.17) allows us to plot the state-plane for time interval Δt_2, shown in Fig. 2.10, which the coordinates of the center of the circle are $(0, V_1 - V_o')$ and the radius is given by (2.18).

$$R_2^2 = \left(V_o' - V_{C1} - V_1\right)^2 + \left(I_1\, z\right)^2 \tag{2.18}$$

Fig. 2.10 State-plane for the time interval Δt_2

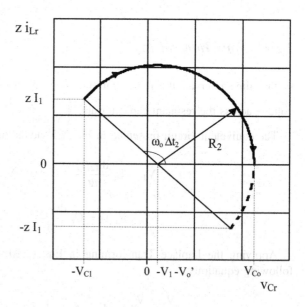

2.3.3 Normalized State Plane for One Half of the Switching Period

The state-plane trajectory for time intervals Δt_1 and Δt_2 of one half of a switching period, normalized with the voltage V_1, is shown in Fig. 2.11, where

$$\overline{i_{Lr}} = \frac{i_{Lr} Z}{V_1}, \; \overline{v_{Cr}} = \frac{v_{Cr}}{V_1}, \; \overline{I_1} = \frac{I_1 Z}{V_1}, \; q = \frac{V'_o}{V_1}, \; q_{Co} = \frac{V_{Co}}{V_1}, \; q_{C1} = \frac{V_{C1}}{V_1}$$

2.3.4 Output Characteristics

The normalized state plane for half of a switching period, shown in Fig. 2.11, yields the following equations:

$$\psi_r + \theta = \pi = \omega_o(\Delta t_1 + \Delta t_2) \tag{2.19}$$

$$\frac{1}{T_s} = \frac{\omega_o}{2\pi} \frac{f_s}{f_o} = \frac{1}{2(\Delta t_1 + \Delta t_2)} \tag{2.20}$$

$$\frac{1}{T_s} = \frac{\mu_o}{\pi} = \frac{1}{\omega_o(\Delta t_1 + \Delta t_2)} = \frac{1}{\psi_r + \theta} \tag{2.21}$$

Fig. 2.11 Normalized state-plane trajectory for one half of a switching period of the series resonant converter, operating in continuous conduction mode

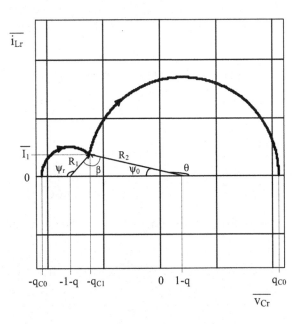

$$\psi_r + \theta = \frac{\pi}{\mu_o} \tag{2.22}$$

and

$$\beta + (\pi - \psi_r) + (\pi - \theta) = \pi \tag{2.23}$$

Substituting (2.22) in (2.23) gives

$$\beta = \psi_r + \theta - \pi = \frac{\pi}{\mu_o} - \pi \tag{2.24}$$

The radios R_1 and R_2 are given by Eqs. (2.25) and (2.26).

$$R_1 = q_{Co} - 1 - q \tag{2.25}$$

$$R_2 = q_{Co} - 1 + q \tag{2.26}$$

In one half of the switching period the capacitor charge (Q) is:

$$Q = C_r 2V_{Co} = \int_0^{T_s/2} i_{Lr}(t)\, dt \tag{2.27}$$

The average load current is

$$I_o' = \frac{1}{T_s/2} \int_0^{T_s/2} i_{Lr}(t)\, dt \tag{2.28}$$

Substituting (2.28) in (2.27), we find:

$$Q = 2C_r V_{Co} = \frac{T_s}{2} I_o' \tag{2.29}$$

and

$$V_{Co} = \frac{I_o'}{4f_s C_r} \tag{2.30}$$

Normalizing V_{Co} in Eq. (2.30) in relation to V_1 gives:

$$q_{Co} = \frac{V_{Co}}{V_1} = \frac{I_o'}{4f_s C_r V_1} \tag{2.31}$$

The normalized average output current is:

$$\overline{I}_o' = \frac{zI_o'}{V_1} \tag{2.32}$$

Substitution of (2.32) into (2.31) yields

$$q_{Co} = \frac{V_{Co}}{V_1} = \frac{\overline{I}_o' V_1}{z} \times \frac{1}{4f_s C_r V_1} = \frac{\overline{I}_o'}{2} \times \frac{\pi}{2\pi f_s C_r z} = \frac{\overline{I}_o'}{2} \times \frac{\pi}{\mu_o} \tag{2.33}$$

Substituting (2.22) in (2.25) and (2.26) we obtain

$$R_1 = \frac{\overline{I}_o'}{2} \times \frac{\pi}{\mu_o} - 1 - q \tag{2.34}$$

$$R_2 = \frac{\overline{I}_o'}{2} \times \frac{\pi}{\mu_o} - 1 + q \tag{2.35}$$

Figure 2.12 shows a detail of the state plane for time intervals Δt_1 and Δt_2, which will be used to find the steady state operation of the series resonant converter. Using the law of cosines, we have

$$2^2 = R_1^2 + R_2^2 - 2R_1R_2 \cos \beta \tag{2.36}$$

Substituting (2.24), (2.34) and (2.35) in (2.36) gives

$$\begin{aligned}
4 = &\left(\frac{\overline{I}_o'}{2} \times \frac{\pi}{\mu_o} - 1 - q\right)^2 + \left(\frac{\overline{I}_o'}{2} \times \frac{\pi}{\mu_o} - 1 + q\right)^2 \\
&- 2\left(\frac{\overline{I}_o'}{2} \times \frac{\pi}{\mu_o} - 1 - q\right)\left(\frac{\overline{I}_o'}{2} \times \frac{\pi}{\mu_o} - 1 + q\right) \cos\left(\frac{\pi}{\mu_o} - \pi\right)
\end{aligned} \tag{2.37}$$

Fig. 2.12 A detail of the state plane for time intervals Δt_1 and Δt_2

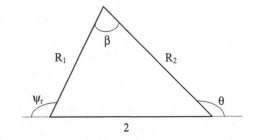

By doing $\rho = \pi/\mu_o$ and knowing that $\cos(\rho - \pi) = -\cos(\rho)$ yields

$$4 = \left(\frac{\overline{I'_o}}{2} \times \rho - 1 - q\right)^2 + \left(\frac{\overline{I'_o}}{2} \times \rho - 1 + q\right)^2$$

$$- 2\left(\frac{\overline{I'_o}}{2} \times \rho - 1 - q\right)\left(\frac{\overline{I'_o}}{2} \times \rho - 1 + q\right)\cos(\rho - \pi) \tag{2.38}$$

After appropriate algebraic manipulation of Eq. (2.38) we find

$$2\left(\frac{\overline{I'_o}\,\rho}{2} - 1\right)^2 + 2q^2 + 2\left[\left(\frac{\overline{I'_o}\,\rho}{2} - 1\right)^2 - q^2\right]\cos(\rho) - 4 = 0 \tag{2.39}$$

Rearranging terms gives

$$2q^2[1 - \cos(\rho)] = 4 - 2\left(\frac{\overline{I'_o}\,\rho}{2} - 1\right)^2[1 + \cos(\rho)] \tag{2.40}$$

Dividing Eq. (2.40) by 4 yields

$$q^2\left(\frac{1 - \cos(\rho)}{2}\right) = 1 - \left(\frac{\overline{I'_o}\,\rho}{2} - 1\right)^2\left(\frac{1 + \cos(\rho)}{2}\right) \tag{2.41}$$

Doing $\frac{1-\cos(\rho)}{2} = \sin^2\left(\frac{\rho}{2}\right)$ and $\frac{1+\cos(\rho)}{2} = \cos^2\left(\frac{\rho}{2}\right)$ in Eq. (2.41) gives

$$q^2 \sin^2\left(\frac{\rho}{2}\right) = 1 - \left(\frac{\overline{I'_o}\,\rho}{2} - 1\right)^2\cos^2\left(\frac{\rho}{2}\right) \tag{2.42}$$

Solving the algebraic Eq. (2.42) for the static gain q yields

$$q = \sqrt{\frac{1 - \left(\frac{\overline{I'_o}\,\rho}{2} - 1\right)^2\cos^2\left(\frac{\rho}{2}\right)}{\sin^2\left(\frac{\rho}{2}\right)}} \tag{2.43}$$

Fig. 2.13 Output
characteristics of the series
resonant converter operating
in continuous current mode

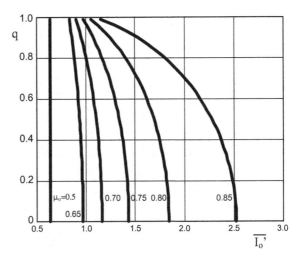

Substituting $\rho = \frac{\pi}{\mu_o}$ in Eq. (2.43) gives

$$q = \sqrt{\frac{1 - \left(\frac{\overline{I_o}\pi}{2\mu_o} - 1\right)^2 \cos^2\left(\frac{\pi}{2\mu_o}\right)}{\sin^2\left(\frac{\pi}{2\mu_o}\right)}} \tag{2.44}$$

Equation (2.44) represents the static gain of the series resonant converter, operating in continuous current mode, as a function of the normalized output current, having μ_o as a parameter. The generated output characteristics are plotted in Fig. 2.13. As it can be noticed, the series resonant converter has a current source characteristic that can be beneficial for protection against overload or short circuit.

2.4   Analysis of the Series Resonant Converter Based in the First Harmonic Approximation

The equivalent circuit of the series resonant converter, operating in continuous conduction mode, with parameters of the secondary side reflected to the primary side of the transformer, is shown in Fig. 2.14.

The transformer is considered ideal and its magnetizing current is null. The amplitude of the fundamental components of the rectangular voltage v_{ab} and v_{cb} are given by the expressions (2.45) and (2.46), respectively.

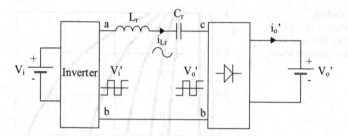

Fig. 2.14 Equivalent circuit of the series resonant converter

Fig. 2.15 Equivalent circuit
for the voltage and current
fundamental components

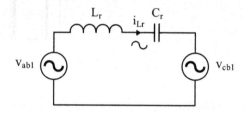

$$V_{ab1} = \frac{4}{\pi} V_1 \tag{2.45}$$

$$V_{cb1} = \frac{4}{\pi} V'_o \tag{2.46}$$

The equivalent circuit of the first harmonic is shown in Fig. 2.15.

Defining the absolute values of x_{Lr} and x_{Cr} by the expressions (2.47) and (2.48), respectively,

$$|x_{Lr}| = L_r\omega_s = 2\pi f_s L_r \tag{2.47}$$

$$|x_{Cr}| = \frac{1}{C_r\omega_s} = \frac{1}{2\pi f_s C_r} \tag{2.48}$$

Hence, the absolute value of the equivalent reactance is given by expression (2.49).

$$|x| = |x_{Cr}| - |x_{Lr}| \tag{2.49}$$

Therefore,

$$|x| = \frac{1}{C_r\omega_s} - L_r\omega_s \tag{2.50}$$

Fig. 2.16 Phasor diagram of
the equivalent circuit shown
in Fig. 2.15

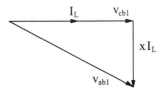

The diode rectifying bridge forces the current i_L to be in phase with the voltage v_{cb1}.

We also know that when the converter operates with $f_s < f_o$, the current i_L leads the voltage v_{ab1} by 90°.

Therefore, in steady state operation, we can represent the voltage and current by the phasor diagram found in Fig. 2.16.

where I_L is the amplitude of the current I_L, considered sinusoidal.

From the phasor diagram, we can write:

$$v_{ab1}^2 = v_{ac1}^2 + (|x|I_L)^2 \tag{2.51}$$

Hence,

$$v_{ac1}^2 = v_{ab1}^2 - (|x|I_L)^2 \tag{2.52}$$

Substituting Eqs. (2.45) and (2.46) into (2.52), we obtain:

$$\left(\frac{4}{\pi}V_o'\right)^2 = \left(\frac{4}{\pi}V_1\right)^2 + (|x|I_L)^2 \tag{2.53}$$

Thus,

$$\left(\frac{\frac{4}{\pi}V_o'}{\frac{4}{\pi}V_1}\right)^2 = 1 - \left(\frac{|x|I_L}{\frac{4}{\pi}V_1}\right)^2 \tag{2.54}$$

The static gain was already defined by

$$q = \frac{V_o'}{V_1} \tag{2.55}$$

Thus,

$$q^2 = 1 - \left(\frac{|x|I_L}{\frac{4}{\pi}V_1}\right)^2 \tag{2.56}$$

or yet,

$$q = \sqrt{1 - \left(\frac{|x| I_L}{\frac{4}{\pi} V_1} \right)^2}$$ (2.57)

Figure 2.17 shows the currents at the diode rectifier input and output.
The average value of the rectified current is:

$$I'_o = \frac{2}{\pi} I_{Lr}$$ (2.58)

Thus,

$$I_{Lr} = \frac{\pi}{2} I'_o$$ (2.59)

Replacing the expression (2.59) in (2.57), we get:

$$q = \sqrt{1 - \left(\frac{|x| \pi I'_o}{2 \frac{4}{\pi} V_1} \right)^2}$$ (2.60)

After the appropriate simplifications, we find:

$$q = \sqrt{1 - \left(\frac{x I'_o \pi^2}{V_1 \, 8} \right)^2}$$ (2.61)

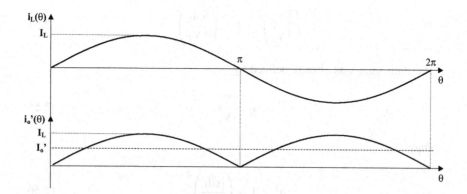

Fig. 2.17 Current at the diode rectifier input and output

The expression (2.62) of the normalized current was already defined by

$$\overline{I}_o' = \frac{z}{V_1} I_o'$$
(2.62)

Thus,

$$I_o' = \frac{V_1}{z} \overline{I}_o'$$
(2.63)

Or,

$$\left(\frac{|x| I_o'}{V_1}\right)^2 = \left(\frac{|x| \overline{I}_o'}{z}\right)^2$$
(2.64)

But,

$$\frac{|x|}{z} = \frac{1}{\sqrt{\frac{L_r}{C_r}}} \left(\frac{1}{\omega_s C_r} - \omega_s L_r\right)$$
(2.65)

With:

$$\omega_o = \frac{1}{\sqrt{L_r C_r}}$$
(2.66)

With the appropriate algebraic manipulation of expressions (2.65) and (2.66), we find:

$$\left(\frac{|x| I_o' \pi^2}{V_1 \, 8}\right)^2 = \left(\frac{\pi^2}{8}\right)^2 \left(\frac{\omega_o}{\omega_s} - \frac{\omega_s}{\omega_o}\right)^2 \overline{I}_o'^2$$
(2.67)

Substituting Eq. (2.55) into (2.61), we find:

$$q = \sqrt{1 - \left[\frac{\pi^2}{8} \left(\frac{\omega_o}{\omega_s} - \frac{\omega_s}{\omega_o}\right) \overline{I}_o'\right]^2}$$
(2.68)

We know that

$$\frac{\omega_o}{\omega_s} = \frac{f_o}{f_s}$$
(2.69)

Therefore, we can write:

$$q = \sqrt{1 - \frac{\pi^4}{64}\left(\frac{f_o}{f_s} - \frac{f_s}{f_o}\right)^2 \overline{I}_o^{-2}}$$

(2.70)

The normalized switching frequency is defined by the expression (2.71):

$$\mu_o = \frac{f_o}{f_s}$$

(2.71)

Substitution of Eq. (2.71) in Eq. (2.70) yields

$$q = \sqrt{1 - \frac{\pi^4}{64}\left(\frac{1}{\mu_o} - \mu_o\right)^2 \overline{I}_o^{-2}}$$

(2.72)

Equation (2.72) represents the output characteristics of the series resonant converter, obtained for voltage and current fundamental components, considering $f_o/2 \leq f_s \leq f_o$.

The output characteristics given by Eq. (2.72) are plotted in Fig. 2.18. The percentage error introduced by the analysis based on the first harmonic approximation is plotted in Fig. 2.19.

Fig. 2.18 Output characteristics of the series resonant converter obtained for the first harmonic approximation

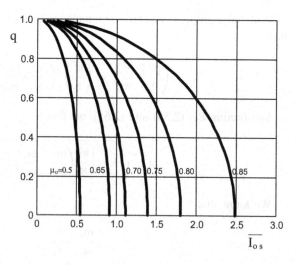

Fig. 2.19 Percentage error as a function of the static gain, taking μ_o as a parameter

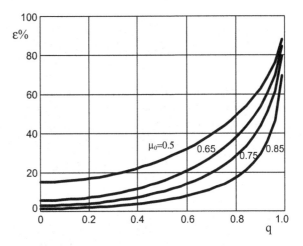

2.5   Simplified Numerical Example for Calculation of L_r and C_r

In this section is presented a numerical example, using the equations deduced in previous sections, to calculate the parameters L_r and C_r of the resonant tank. The specifications are given in Table 2.1.

A normalized switching frequency $\mu_o = 0.87$ is chosen to provide a wide load range operation in continuous conduction mode. The maximum normalized output current is $\overline{I'_{o\,max}} = 2.5$ for a static gain q = 0.6.

The output voltage referred to the transformer primary side is:

$$V'_o = \frac{V_i}{2}\,q = \frac{400}{2} \times 0.6 = 120 \text{ V}$$

The transformer turns ratio (n) and the average output current referred to the primary side of the transformer (I_o') are:

$$\frac{N_1}{N_2} = \frac{V'_o}{V_o} = \frac{120}{50} = 2.4$$

Table 2.1 Design specifications

Input DC voltage (V_i)	400 V
Output DC voltage (V_o)	50 V
Output DC current (I_o)	10 A
Output nominal power (P_o)	500 W
Minimum output power ($P_{o\,min}$)	50 W
Maximum switching frequency ($f_{s\,max}$)	40 kHz

and

$$I'_o = V_o \frac{N_1}{N_2} = 10 \times \frac{1}{2.4} = 4.17 \text{ A}$$

The average output current referred to the primary side of the transformer gives a relation between the inductance of the resonant inductor and the capacitance of the resonant capacitor. Hence,

$$\overline{I'_o} = \frac{I'_o \sqrt{\frac{L_r}{C_r}}}{V_1}$$

and

$$\frac{L_r}{C_r} = \left(\frac{\overline{I_o} V_1}{I_o}\right)^2 = \left(\frac{2.5 \times 200}{4.16667}\right)^2 = 14,400$$

The resonant frequency is

$$f_o = \frac{f_{s\,max}}{\mu_o} = \frac{40 \times 10^3}{0.87} = 45,977.01 \text{ Hz}$$

Since $f_o = \frac{1}{2\pi\sqrt{L_r C_r}}$, then $L_r C_r = 11.98276 \times 10^{-12}$.

From these two equations we find $C_r = 29$ nF and $L_r = 413.23$ μH.

For the minimum power $P_{o\,min} = 50$ W, $q = 0.6$ and $\overline{I'_{o\,min}} = 1$, from the output characteristics we have $\mu_o = 0.7$. Hence, the minimum switching frequency is:

$$f_{s\,min} = f_o \mu_o = 45,977 \times 0.7 = 32,183.91 \text{ Hz}$$

The operating region of the output characteristics for the calculated parameters is presented in Fig. 2.20.

The ideal half-bridge series resonant converter, whose equivalent circuit is shown in Fig. 2.21, with the parameters given in Table 2.1 and the resonant tank parameters calculated above, is simulated at nominal and minimum load power.

Figure 2.22 shows the resonant capacitor voltage, the resonant inductor current and the voltage v_{ab} at the nominal power and Fig. 2.23 shows the voltage and current in the switches. As it may be noticed, the switches turn on is dissipative and soft commutation (ZCS) is achieved only at the switches turn off.

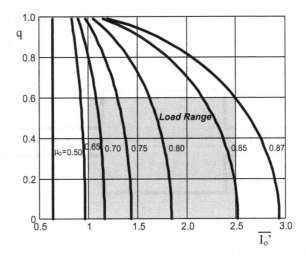

Fig. 2.20 Output characteristics showing the operation region of the numerical example

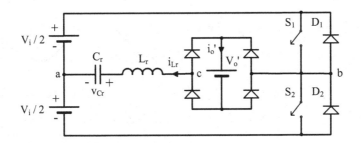

Fig. 2.21 Ideal half-bridge series resonant converter

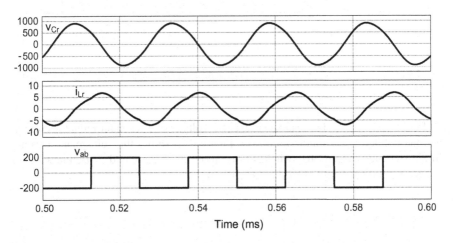

Fig. 2.22 Resonant capacitor voltage, resonant inductor current and voltage v_{ab} at nominal power

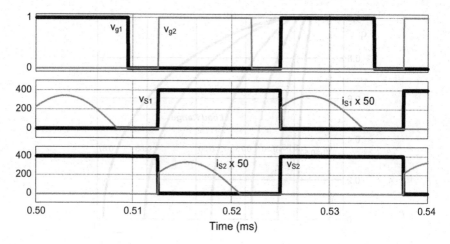

Fig. 2.23 Switches commutation at nominal power: S_1 and S_2 drive signals, voltage and current on the switches

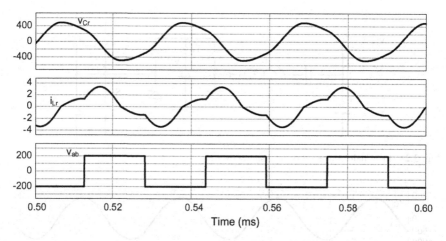

Fig. 2.24 Resonant capacitor voltage, resonant inductor current and voltage v_{ab} at minimum power

   Figure 2.24 shows the resonant capacitor voltage, the resonant inductor current and the voltage v_{ab} at minimum power. Figure 2.25 shows the voltage and current in the switches. Again, the switches turn on are dissipative and soft commutation (ZCS) is achieved only when the switch turns off.

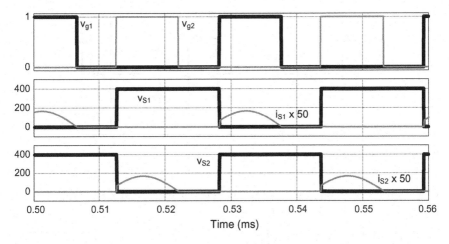

Fig. 2.25 Switches commutation at minimum power: S_1 and S_2 drive signals, voltage and current on the switches

2.6   Operation in Discontinuous Conduction Mode

In DCM the switching frequency range is $0 \leq f_s \leq 0.5 \, f_o$ and in one operation cycle there are six time intervals.

(A)   Time Interval Δt_1 ($t_0 < t < t_1$)

At $t = t_0$, because $i_{Lr} = 0$, the switch S_1 turn on is non-dissipative (ZCS). The inductor current evolves sinusoidally, and the capacitor that in the beginning of this time interval was charged with a negative voltage is discharged. During this time interval the DC bus delivers energy to the load. Figure 2.26 shows the topological state for this interval.

(B)   Time Interval Δt_2 ($t_1 < t < t_2$)

The topological state for this time interval is illustrated in Fig. 2.27. In $t = t_1$ the inductor current reaches zero. The diode D_1 turns on conducting the inductor current that evolves negatively. During this time interval switch S_1 must be turned

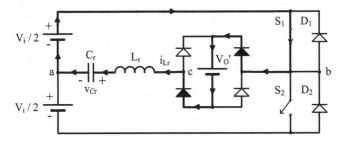

Fig. 2.26 Topological state for time interval Δt_1

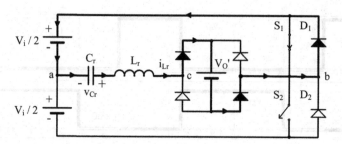

Fig. 2.27 Topological state for time interval Δt_2

off to ensure soft commutation. This time interval ends when the inductor current reaches zero again, at the instant $t = t_2$.

(C) Time Interval Δt_3 ($t_2 < t < t_3$)

During this time interval, shown in Fig. 2.28, both switches are open and the inductor current is null. No power is transferred to the load. This time interval finishes at $t = t_3$, when switch S_2 is turned on.

(D) Time Interval Δt_4 ($t_3 < t < t_4$)

At $t = t_3$ the inductor current is zero and switch S_2 turns on with zero current. The inductor current evolves sinusoidally, and the capacitor is discharged. During this time interval the DC bus delivers energy to the load. Figure 2.29 shows the topological state for this time interval.

(E) Time Interval Δt_5 ($t_4 < t < t_5$)

The topological state for this time interval is presented in Fig. 2.30. At $t = t_4$ the inductor current reaches zero. The diode D_2 starts conducting the inductor current that evolves positively. During this time interval switch S_2 must be turned off to ensure soft commutation. This interval ends when the inductor current reaches zero again.

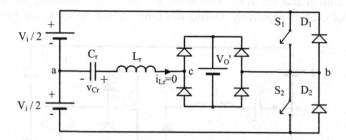

Fig. 2.28 Topological state for time interval Δt_3

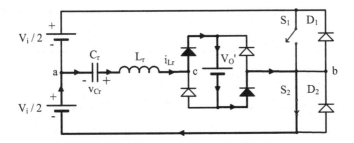

Fig. 2.29 Topological state for time interval Δt_4

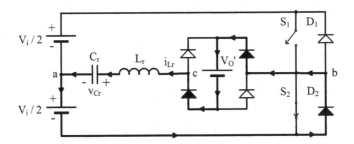

Fig. 2.30 Topological state for time interval Δt_5

(F) Time Interval Δt_6 ($t_5 < t < t_6$)

During this time interval, presented in Fig. 2.31, both switches are open and the inductor current is null. No power is transferred to the load. This interval finishes at $t = t_6$, when switch S_1 is turned on. The main waveforms with the time diagram are shown in Fig. 2.32.

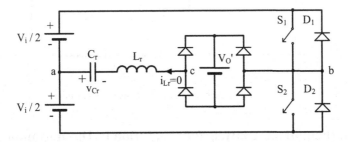

Fig. 2.31 Topological state for time interval Δt_6

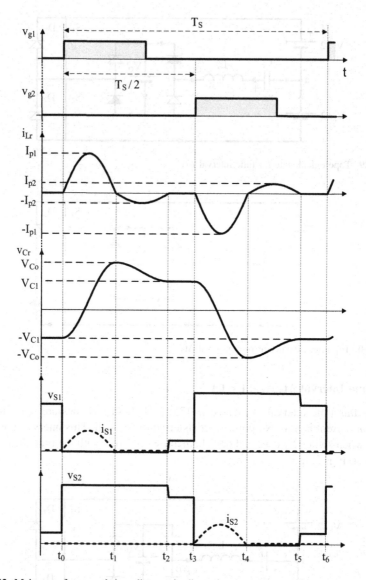

Fig. 2.32 Main waveforms and time diagram in discontinuous current mode

2.7 Mathematical Analysis for Operation in Discontinuous Conduction Mode

In this section the resonant inductor current and the resonant capacitor voltage are obtained. Since the converter is symmetrical, only one half switching period is analyzed.

(A) A Time Interval Δt_1

In this time interval, the initial current in the resonant inductor and voltage across the resonant capacitor are: $\begin{cases} i_{Lr}(t_0) = 0 \\ v_{Cr}(t_0) = -V_{C0} \end{cases}$

The equivalent circuit shown in Fig. 2.26 yields the following equations:

$$\frac{V_i}{2} = L_r \frac{di_{Lr}(t)}{dt} + v_{Cr}(t) + V'_o \tag{2.73}$$

$$i_{Lr}(t) = C_r \frac{dv_{Cr}(t)}{dt} \tag{2.74}$$

Applying the Laplace Transform we find

$$\frac{(V_i/2) - V'_o}{s} = sL_r I_{Lr}(s) + v_{Cr}(s) \tag{2.75}$$

$$I_{Lr}(s) = sC_r v_{Cr}(s) + C_r V_{C1} \tag{2.76}$$

Applying the inverse Laplace Transform in Eqs. (2.75) and (2.76) and rearranging terms, we have

$$v_{Cr}(t) = -\left(V_1 - V'_o + V_{C1}\right) \cos\left(\omega_o t\right) + V_1 - V'_o \tag{2.77}$$

$$i_{Lr}(t) \, z = \left(V_1 - V'_o + V_{C1}\right) \operatorname{sen}\left(\omega_o t\right) \tag{2.78}$$

This time interval ends when the inductor current reaches zero. Considering $i_{Lr}(t) = 0$ in Eq. (2.78) gives

$$\left(V_1 - V'_o + V_{C1}\right) \operatorname{sen}(\omega_o \Delta t_1) = 0 \tag{2.79}$$

Hence,

$$\Delta t_1 = \frac{\pi}{\omega_o} \tag{2.80}$$

Defining

$$z_1(t) = v_{Cr}(t) + j \, z \, i_{Lr}(t) \tag{2.81}$$

Substituting (2.77) and (2.78) in (2.81) leads to

$$z_1(t) = V_1 - V'_o - \left(V_1 - V'_o + V_{C1}\right) \cos(\omega_o t) + j \left(V_1 - V'_o + V_{C1}\right) \operatorname{sen}(\omega_o t) \tag{2.82}$$

Fig. 2.33 State-plane trajectory for time interval Δt_1, in discontinuous conduction mode

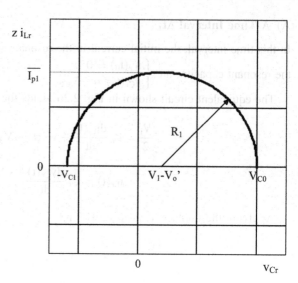

Hence,

$$z_1(t) = V_1 - V_o' - \left(V_1 - V_o' + V_{C1}\right) e^{-j\omega_o t} \qquad (2.83)$$

Equation (2.83) gives the state-plane trajectory for time interval Δt_1, shown in Fig. 2.33. It is a circle whose center coordinates is $(0, V_1 - V_o')$ and radius is

$$R_1 = V_{C1} + V_1 - V_o' \qquad (2.84)$$

(B) Time Interval Δt_2

In this time interval, the initial current in the resonant inductor and the voltage across the resonant capacitor are: $\begin{cases} i_{Lr}(t_1) = 0 \\ v_{Cr}(t_1) = V_{C0} \end{cases}$.

The equivalent circuit shown in Fig. 2.27 yields the following equations:

$$V_1 = -L_r \frac{di_{Lr}(t)}{dt} + v_{Cr}(t) - V_o' \qquad (2.85)$$

$$i_{Lr}(t) = -C_r \frac{dv_{Cr}(t)}{dt} \qquad (2.86)$$

Applying the Laplace Transform, we find:

$$\frac{V_1 + V_o'}{s} = -sL_r i_{Lr}(s) + v_{Cr}(s) \qquad (2.87)$$

$$i_{Lr}(s) = -sC_r v_{Cr}(s) + C_r V_{CO} \tag{2.88}$$

Applying the inverse Laplace Transform in Eqs. (2.87) and (2.88) and rearranging, we find

$$v_{Cr}(t) = -\left(V_1 + V'_o - V_{CO}\right)\cos(\omega_o t) + V_1 + V'_o \tag{2.89}$$

$$i_{Lr}(t)\, z = \left(V_1 + V'_o - V_{CO}\right)\mathrm{sen}(\omega_o t) \tag{2.90}$$

This time interval ends when the inductor current reaches zero. Considering $i_{Lr}(t) = 0$ in Eq. (2.90) yields

$$\left(V_1 + V'_o - V_{CO}\right)\mathrm{sen}(\omega_o \Delta t_2) = 0 \tag{2.91}$$

Thus,

$$\Delta t_2 = \frac{\pi}{\omega_o} \tag{2.92}$$

Defining the variable $z_2(t)$ as follows:

$$z_2(t) = v_{Cr}(t) + j\, z\, i_{Lr}(t) \tag{2.93}$$

Substituting (2.89) and (2.90) in (2.93) yields

$$z_2(t) = V_1 + V'_o - \left(V_1 + V'_o - V_{CO}\right)\cos(\omega_o t) + j\left(V_1 + V'_o - V_{CO}\right)\mathrm{sen}(\omega_o t) \tag{2.94}$$

Hence,

$$z_2(t) = V_1 + V'_o - \left(V_1 + V'_o - V_{CO}\right)e^{-j\,\omega_o t} \tag{2.95}$$

Equation (2.95) allows us to plot the state-plane trajectory for time interval Δt_2 as shown in Fig. 2.34. The coordinates of the center of the circle are $(0, V_1 + V'_o)$ and the radius is

$$R_2 = \left[V_{CO} - \left(V_1 + V'_o\right)\right] \tag{2.96}$$

(C) Initial Conditions

The complete state-plane trajectory of a half switching period is shown in Fig. 2.35. By inspection of the figure, we can write Eqs. (2.97) and (2.98).

Fig. 2.34 State-plane trajectory for time interval Δt_2, in discontinuous conduction mode

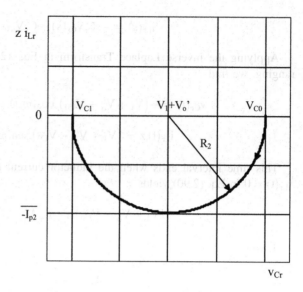

Fig. 2.35 State-plane trajectory for half-period, in discontinuous conduction mode

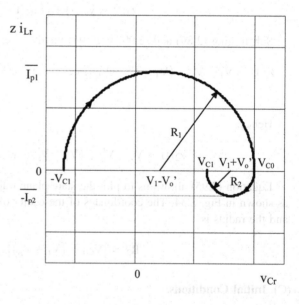

$$V_{C1} = V_{C0} - 2R_2 \tag{2.97}$$

$$V_{C0} = R_1 + \left(V_1 - V'_o\right) \tag{2.98}$$

Substitution of Eqs. (2.96) in (2.97) and Eq. (2.84) in (2.98) gives:

$$V_{C1} = -V_{C0} + 2\left(V_1 + V'_o\right) \tag{2.99}$$

$$V_{C0} = V_{C1} + 2\left(V_1 - V'_o\right) \tag{2.100}$$

Substituting (2.99) in (2.100) we find:

$$V_{C1} = 2\,V'_o \tag{2.101}$$

$$V_{C0} = 2\,V_1 \tag{2.102}$$

The radius R_1 and R_2 are determined by:

$$R_1 = V_1 + V'_o \tag{2.103}$$

$$R_2 = V_1 - V'_o \tag{2.104}$$

The peak currents I_{p1} and I_{p2} are given by Eqs. (2.105) and (2.106), respectively.

$$\overline{I_{p1}} = I_{p1}\,z = R_1 = V_1 + V'_o \tag{2.105}$$

$$\overline{I_{p2}} = I_{p2}\,z = R_2 = V_1 - V'_o \tag{2.106}$$

(D) Output Characteristics

Figure 2.36 shows a detail of the power stage diagram of the series resonant converter, whose currents at the input and output terminals of the diode rectifier are shown in Fig. 2.37.

The areas A_1 and A_2 are determined by Eqs. (2.107) and (2.108), respectively.

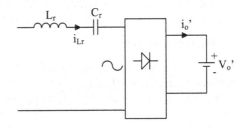

Fig. 2.36 Detail of the power stage diagram of the series resonant converter

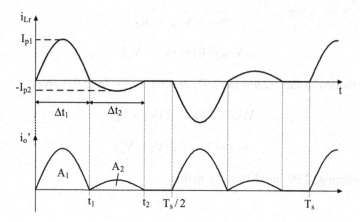

Fig. 2.37 Currents at the input and output terminals of the rectifier

$$A_1 = \frac{1}{\omega_o} \int_0^\pi \left(\frac{V_1 + V_o'}{Z}\right) \text{sen}(\omega_o t)\, dt = \frac{V_1 + V_o'}{\pi f_o Z} \qquad (2.107)$$

$$A_2 = \frac{1}{\omega_o} \int_0^\pi \frac{V_1 - V_o'}{Z} \text{sen}(\omega_o t)\, dt = \frac{V_1 - V_o'}{\pi f_o Z} \qquad (2.108)$$

The average output current, referred to the transformer primary side, is obtained by adding (2.107) and (2.108). Hence,

$$I_{oD}' = \frac{2(A_1 + A_2)}{T_s} = \frac{4}{\pi} \frac{V_1}{z} \frac{f_s}{f_o} \qquad (2.109)$$

Thus, the normalized current is

$$\overline{I_{oD}'} = \frac{I_{oD}' z}{V_1} = \frac{4}{\pi} \frac{f_s}{f_o} \qquad (2.110)$$

The output characteristics or the static gain as a function of the normalized output current, having μ_o as a parameter, are plotted in Fig. 2.38. It should be noted that the converter, in this operation mode, has current source characteristic, so the load current does not depend of the load voltage.

(E) DCM Constraints

The discontinuous conduction mode described in Sect. 2.6 is only possible if three constraints are observed.

The first one is given by Eq. (2.111).

Fig. 2.38 Output characteristics in discontinuous current mode

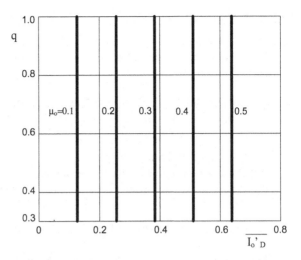

$$\omega_o(\Delta t_1 + \Delta t_2) \leq \pi \tag{2.111}$$

Hence, substituting (2.80) and (2.92) in (2.111) we find

$$\frac{f_s}{f_o} \leq 0.5 \tag{2.112}$$

Substitution of Eq. (2.112) to (2.110) gives the second constraint, given by Eq. (2.113).

$$\overline{I'_{oD}} \leq \frac{2}{\pi} \tag{2.113}$$

The third constraint, obtained by inspection of the state-plane trajectory of Fig. 2.34 is:

$$V_{C1} \geq V_1 - V'_o \tag{2.114}$$

Substituting (2.101) in (2.114), we find:

$$V'_o \geq \frac{V_1}{3} \tag{2.115}$$

2.8 Problems

(1) The specification of the full-bridge series resonant converter shown in Fig. 2.39
    are:

$$V_i = 400\,V \qquad V_o = 120\,V \qquad N_s/N_p = 1 \qquad R_o = 20\,\Omega$$
$$f_s = 30 \times 10^3\,Hz \qquad C_r = 52.08\,nF \qquad L_r = 48.63\,\mu H$$

   (a) Calculate the resonant frequency.
   (b) Calculate the output voltage and the output average current.
   (c) Calculate the resonant capacitor peak voltage and the switches peak
       current.

   Answers: (a) f_o = 100 kHz, (b) V_o = 100 V; I_o = 5 A and (c) V_{CO} = 800 V;
   I_{p1} = 16.36 A

(2) For exercise 1 calculate the switching frequency to ensure that the output
    voltage is 160 V.
    Answer: f_s = 48 kHz

(3) The full-bridge series resonant converter shown in Fig. 2.39 operates in CCM
    with a switching frequency of 80 kHz. Calculate the output voltage.
    Answer: V_o = 344 V

(4) Find the expression of the output short-circuit current for a series resonant
    converter operating in CCM.
    Answer: $\left[I_{o\_SC} = \frac{V_1}{Z} \times \frac{2\mu_o}{\pi} \times \left(1 + \frac{1}{\cos(\pi/2\,\mu_o)} \right) \right]$

(5) For exercise 4 calculate the output short circuit current.
    Answer: I_{o_SC} = 24 A

(6) Find the expression for the resonant capacitor peak voltage of the half-bridge
    series resonant converter operating with the output short-circuited.
    Answer: $V_{Co\,SC} = V_1 \times \left(1 + \frac{1}{\left| \cos\left(\frac{\pi}{2\mu_o} \right) \right|} \right)$

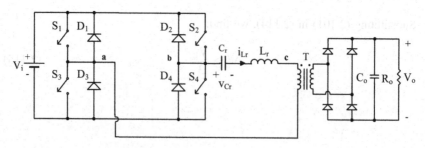

Fig. 2.39 Full-bridge series resonant converter

(7) Calculate the minimum output resistance ($R_{o\ min}$) to ensure that a half-bridge series resonant converter operates in discontinuous conduction mode. Consider the following specifications:

$$V_i = 400\,V \qquad C_r = 300\,nF \quad L_r = 40\,\mu H$$
$$f_s = 20 \times 10^3\,Hz \quad N_s/N_p = 1$$

Answer: $R_{o\ min} = 15.625\ \Omega$

References

1. Schwarz, F.C.: An improved method of resonant current pulse modulation for power converters. IEEE Trans. Ind. Electron. Control Instrum. **IECI-23**(2), 133–141 (1976)
2. Schwarz, F.C., Klaassens, B.J.: A 95-percent efficient 1-kW DC converter with an internal frequency of 50 kHz. IEEE Trans. Ind. Electron. Control Instrum. **IECI-24**(4), 326–333 (1978)
3. Witulski, A.F., Erickson, R.W.: Steady-state analysis of the series resonant converter. IEEE Trans. Aerosp. Electron. Syst. **AES-21**(6), 791–799 (1985)

Chapter 3
Half Bridge Capacitor Voltage-Clamped Series Resonant Converter

Nomenclature

V_i	Input DC voltage
V_o	Output DC voltage
V_1	Half of the DC input voltage
P_o	Output power
$P_{o\ min}$	Minimum output power
$P_{o\ max}$	Maximum output power
C_o	Output filter capacitor
R_o	Output load resistor
μ_o	Normalized switching frequency
q	Static gain
D	Duty cycle
f_s	Switching frequency
$f_{s\ max}$	Maximum switching frequency
$f_{s\ min}$	Minimum switching frequency
T_s	Switching period
t_d	Dead time
T	Transformer
n	Transformer turns ratio
N_1 and N_2	Transformer primary and secondary winding turns
V'_o	Output DC voltage referred to the transformer primary side
i_o	Output current
i'_o	Output current referred to the transformer primary side
$I'_o\left(\overline{I'_o}\right)$	Average output current referred to the transformer primary side and its normalized value
S_1 and S_2	Switches
v_{g1} and v_{g2}	Switches gate signals
D_{C1} and D_{C2}	Clamping diodes
C_r	Resonant capacitor
L_r	Resonant inductor (may include the transformer leakage inductance)
i_{Lr}	Resonant inductor current

© Springer International Publishing AG, part of Springer Nature 2019
I. Barbi and F. Pöttker, *Soft Commutation Isolated DC-DC Converters*,
Power Systems, https://doi.org/10.1007/978-3-319-96178-1_3

I_{Lr} $\left(\overline{I_{Lr}}\right)$	Resonant inductor peak current and its normalized value, commutation current for S_1 and S_3
ω_o	Resonant frequency
z	Characteristic impedance
z_1 and z_2	State plane impedance
I_1 $\left(\overline{I_1}\right)$	Inductor current at the end of the first and fourth step of operation and its normalized value
v_{ab}	Full bridge ac voltage, between points "a" and "b"
v_{S1} and v_{S2}	Voltage across switch S_1 and S_2
i_{S1} and i_{S2}	Current in the switch S_1 and S_2
Δt_1	Time interval of the first step of operation in (t_1-t_0)
Δt_2	Time interval of the second step of operation (t_2-t_1)
Δt_3	Time interval of the third step of operation in (t_3-t_2)
Δt_4	Time interval of the fourth step of operation (t_4-t_3)
Δt_5	Time interval of the fifth step of operation in (t_5-t_4)
Δt_6	Time interval of the sixth step of operation (t_6-t_5)
$I_{S\,RMS}$ $\left(\overline{I_{S\,RMS}}\right)$	Switches S_1 and S_2 RMS currents and its normalized value
I_{DC} $\left(\overline{I_{DC}}\right)$	Diodes D_{C1} and D_{C2} average currents and its normalized value

3.1   Introduction

The half bridge capacitor voltage-clamped series resonant converter [1], presented in Fig. 3.1, has two clamping diodes compared to the series resonant converter presented in Chap. 2. The power is transferred to the load is also controlled by the switching frequency. The resonant capacitor voltage is clamped with half of the DC Bus voltage, resulting in two linear time intervals in the switching period. The converter operates in discontinuous conduction mode (DCM) in a wide switching

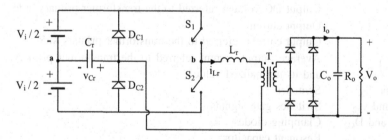

Fig. 3.1 Half bridge capacitor voltage-clamped series resonant converter

frequency range, up to the resonant frequency. The operation in DCM ensures a zero-current switching (ZCS) at the semiconductors when it is turned on and off. The output characteristic is not an ideal current source as it is in the series resonant converter, so short circuit protection is not inherent to this topology. The output power depends linearly on the switching frequency and does not depend on the load. The schematics of the HB CVC-SRC is shown in Fig. 3.1.

In this chapter, principle of operation, detailed analysis in discontinuous conduction mode (DCM), as well as a design example are presented.

3.2 Circuit Operation

In this section, the converter shown in Fig. 3.2 is analyzed. The following assumptions are made:

- all components are considered ideal;
- the converter is in steady-state operation;
- the output filter is represented as an DC voltage V'_o, whose value is the output voltage referred to the primary winding of the transformer;
- current that flows through the switches is unidirectional, i.e. the switches allow the current to flow only in the direction of the arrow;
- the converter is controlled by frequency modulation, the switches S_1 and S_2 are gated with 50% duty cycle, and dead time is not considered.

The converter is analyzed in DCM. In this conduction mode the switches turn on and off are non-dissipative, with zero current (ZCS).

(A) Time Interval Δt_1 ($t_0 \leq t \leq t_1$)

This time interval, shown in Fig. 3.3, starts at $t = t_0$, when the switch S_1 is gated on. Prior to $t = t_0$, the capacitor voltage is $-V_i/2$ and the inductor current is zero, so switch S_1 turns on softly, with zero current (ZCS). During this time interval, both the capacitor voltage and inductor current evolve in a resonant way.

Fig. 3.2 Half bridge CVC-SRC converter

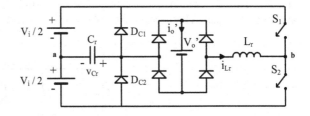

Fig. 3.3 Topological state
for time interval Δt_1

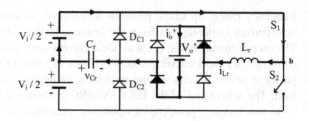

Fig. 3.4 Topological state
for time interval Δt_2

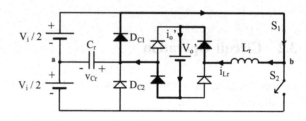

Fig. 3.5 Topological state
for time interval Δt_3

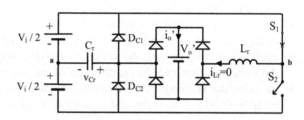

(B) Time Interval Δt_2 ($t_1 \leq t \leq t_2$)

Time interval Δt_2, presented in Fig. 3.4, begins at the instant $t = t_1$, when the
capacitor voltage reaches $V_i/2$ and the clamping diode D_{C1} starts conducting the
inductor current i_{Lr}. The capacitor voltage is clamped with $V_i/2$ and the inductor
current decreases linearly and eventually reaches zero at the instant $t = t_2$.

(C) Time Interval Δt_3 ($t_2 \leq t \leq t_3$)

During this time interval switch S_2 is not gated on yet, the capacitor voltage is $V_i/2$,
and the inductor current is zero. No power is transferred from the input voltage
source to the load. The converter topological state is illustrated in Fig. 3.5.

Fig. 3.6 Topological state
for time interval Δt_4

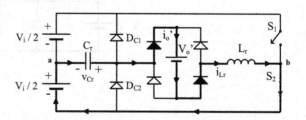

Fig. 3.7 Topological state
for time interval Δt_5

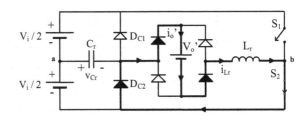

Fig. 3.8 Topological state
for time interval Δt_6

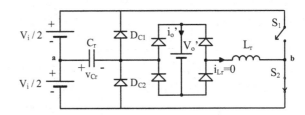

(D) Time Interval Δt_4 ($t_3 \leq t \leq t_4$)

In the instant $t = t_3$, that corresponds to half of the switching period, switch S_2 is gated on with zero current (ZCS). The capacitor voltage and inductor current evolve in a resonant way until $v_{Cr}(t) = -V_i/2$ and $i_{Lr}(t) = -I_1$, at the instant $t = t_4$. The equivalent circuit for this time interval may be observed at Fig. 3.6.

(E) Time Interval Δt_5 ($t_4 \leq t \leq t_5$)

Time interval Δt_5 begins when the capacitor voltage reaches $-V_i/2$, turning on the clamping diode D_{C2}. The capacitor voltage is clamped at $-V_i/2$ and the inductor current decreases linearly until it reaches zero. The topological state for this interval is presented in Fig. 3.7.

(F) Time Interval Δt_6 ($t_5 \leq t \leq t_6$)

During this time interval switch S_1 is not gated on yet, the capacitor voltage is clamped with $-V_i/2$ and the inductor current is zero. No power is transferred from the input voltage source to the load. The topological state is shown in Fig. 3.8.

The main waveforms and the time diagram are shown in Fig. 3.9.

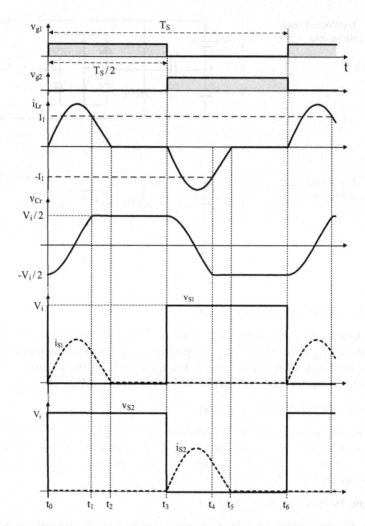

Fig. 3.9 Main waveforms and time diagram of the half bridge CVC-SRC

3.3 Mathematical Analysis

In this section, the resonant capacitor voltage and resonant inductor current expressions are obtained likewise the DCM operations constraints, the output characteristics and output power. The switches RMS current and clamping the diodes average current. Since the converter is symmetrical, only half of the switching period is analyzed.

3.3.1 Time Interval Δt_1

In time interval Δt_1, the DC bus delivers energy to the resonant tank and the load. The initial current in the resonant inductor and the initial voltage across the resonant capacitor are: $\begin{cases} i_{Lr}(t_0) = 0 \\ v_{Cr}(t_0) = -V_i/2 \end{cases}$

From the equivalent circuit shown in Fig. 3.3, we have

$$\frac{V_i}{2} = L_r \frac{di_{Lr}(t)}{dt} + v_{Cr}(t) + V_o' \tag{3.1}$$

$$i_{Lr}(t) = C_r \frac{dv_{Cr}(t)}{dt} \tag{3.2}$$

Applying the Laplace Transform yields

$$\frac{(V_i/2) - V_o'}{s} = s L_r i_{Lr}(s) + v_{Cr}(s) \tag{3.3}$$

$$i_{Lr}(s) = s C_r v_{Cr}(s) + C_r \frac{V_i}{2} \tag{3.4}$$

$$i_{Lr}(t) z = (2 V_1 - V_o') \operatorname{sen}(\omega_0 t) \tag{3.6}$$

where $V_1 = \frac{V_i}{2}$, $\omega_0 = \frac{1}{\sqrt{L_r C_r}}$, $q = \frac{V_o'}{V_1}$ and $z = \sqrt{\frac{L_r}{C_r}}$.

Normalizing the capacitor voltage and the inductor current with the voltage V_1 gives

$$\overline{v_{Cr}(t)} = \frac{v_{Cr}(t)}{V_1} = -(2 - q) \cos(\omega_0 t) + 1 - q \tag{3.7}$$

$$\overline{i_{Lr}(t)} = \frac{i_{Lr}(t) z}{V_1} = (2 - q) \operatorname{sen}(\omega_0 t) \tag{3.8}$$

Substituting (3.4) in (3.3) and applying the inverse Laplace Transform we find:

$$v_{Cr}(t) = -(2 V_1 - V_o') \cos(\omega_0 t) + V_1 - V_o' \tag{3.5}$$

This time interval finishes when the capacitor voltage reaches V_1 and consequently $v_{Cr}(t) = 1$. Therefore, from Eq. (3.7) we have

$$1 = -(2 - q) \cos(\omega_0 \Delta t_1) + 1 - q \tag{3.9}$$

Thus,

$$\omega_o \, \Delta t_1 = \pi - \cos^{-1}\left(\frac{1}{2-q}\right) \tag{3.10}$$

Time interval Δt_1, shown in (3.10), does not depend on the switching frequency. In the end of this time interval the inductor current is:

$$\overline{I}_1 = \frac{i_{Lr}(t_1)\, z}{V_1} = (2-q)\, \text{sen}\left[\pi - \cos^{-1}\left(\frac{q}{2-q}\right)\right] \tag{3.11}$$

After appropriate algebraic manipulation we find

$$\overline{I}_1 = 2\sqrt{1-q} \tag{3.12}$$

Defining the variable $z_1(t)$ as follows:

$$z_1(t) = v_{Cr}(t) + j\, z\, i_{Lr}(t) \tag{3.13}$$

Substitution of Eqs. (3.7) and (3.8) into Eq. (3.13) yields:

$$z_1(t) = -(2-q)\,\cos(\omega_o t) + 1 - q + j\,(2-q)\,\text{sen}(\omega_o t) \tag{3.14}$$

Hence,

$$z_1(t) = (1-q) - (2-q)\,e^{-j\,\omega_o t} \tag{3.15}$$

Equation (3.15) allows us to plot the state-plane for time interval Δt_1, shown in Fig. 3.10. The state-plane trajectory is a circle with center at the coordinates (0, 1 −q) and radius given by Eq. (3.16).

$$R_1 = 2 - q \tag{3.16}$$

Fig. 3.10 State plane diagram of HB CVC-SRC for an operation cycle

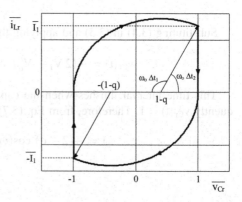

3.3.2 Time Interval Δt_2

In time interval Δt_2, the energy stored in the resonant inductor at the end of the time interval Δt_1 is totally transferred to the load.

The initial current in the resonant inductor and the initial voltage across the resonant capacitor are: $\begin{cases} i_{Lr}(t_0) = I_1 \\ v_{Cr}(t_0) = V_1 \end{cases}$

From the equivalent circuit shown in Fig. 3.4, we have

$$i_{Lr}(t) = I_1 - \frac{V'_o}{L_r}(t - t_1) \tag{3.17}$$

and

$$v_{Cr}(t) = V_1 \tag{3.18}$$

Normalizing (3.17) and (3.18) with V_1 gives

$$\overline{v_{Cr}(t)} = \frac{v_{Cr}(t)}{V_1} = 1 \tag{3.19}$$

and

$$\overline{i_{Lr}(t)} = \frac{i_{Lr}(t)\,z}{V_1} = \overline{I_1} - q\,\omega_o\,(t - t_1) \tag{3.20}$$

This time interval finishes when the inductor current reaches zero. Hence, from Eq. (3.20) we have

$$0 = 2\sqrt{1 - q} - q\,\omega_o\,(\Delta t_2) \tag{3.21}$$

Thus,

$$\omega_o\,\Delta t_2 = \frac{2\sqrt{1 - q}}{q} \tag{3.22}$$

Defining $z_2(t)$ according to Eq. (3.23).

$$z_2(t) = \overline{v_{Cr}(t)} + j\,\overline{i_{Lr}(t)} \tag{3.23}$$

Substitution of Eqs. (3.19) and (3.20) in Eq. (3.23) gives

$$z_2(t) = 1 + j\left[\overline{I_1} - q\,\omega_o\,(t - t_1)\right] \tag{3.24}$$

The state-plane trajectory for this time interval is a line, parallel to $\overline{i_{Lr}}$ axis, as shown in Fig. 3.10.

3.3.3   State Plane Diagram

The state plane trajectory of the HB CVC-SRC is shown in Fig. 3.10, with resonant and linear time intervals, for a complete operation cycle of the converter.

3.3.4   DCM Constrains

In the limit of operation in discontinuous conduction $\Delta t_3 = 0$ and

$$\frac{T_s}{2} = \Delta t_1 + \Delta t_2 \tag{3.25}$$

Thus,

$$\frac{\pi}{f_s/f_o} = \omega_o \left(\Delta t_1 + \Delta t_2 \right) \tag{3.26}$$

Substitution of Eqs. (3.10) and (3.22) into Eq. (3.26) gives:

$$\frac{f_{s\,max}}{f_o} = \frac{\pi}{\pi - \cos^{-1}\left(\frac{q}{2-q} \right) + \frac{2\sqrt{1-q}}{q}} \tag{3.27}$$

Equation (3.27) gives the maximum switching frequency, normalized with the resonant frequency, as a function of the static gain, which is plotted in Fig. 3.11. As the static gain increases, the maximum switching frequency increases too. The limit is with $q = 1$, when $f_{smax} = f_o$. If the maximum switching frequency is not respected, the converter will operate in continuous conduction mode (CCM), and soft commutation will not be achieved when the semiconductors turn on and off.

Fig. 3.11 Normalized maximum switching frequency, as a function of the static gain q

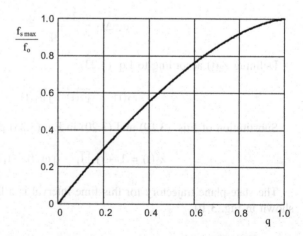

3.3.5 Output Characteristics

The average output current is calculated according to (3.28)

$$\overline{I'_0} = \frac{2}{T_s} \left(\int_{t_0}^{t_1} \overline{i_{Lr}(t)}\, dt + \int_{t_1}^{t_2} \overline{i_{Lr}(t)}\, dt \right) \tag{3.28}$$

Substituting (3.8) and (3.20) in Eq. (3.28) and integrating, gives

$$\overline{I'_0} = \frac{2}{\omega_o T_s} \left[2 + \frac{2(1-q)}{q} \right] \tag{3.29}$$

Rearranging terms we find Eq. (3.30) that represents the static gain of the converter.

$$q = \frac{I'_0 z}{V_1} = \frac{2}{\pi} \times \frac{1}{\overline{I'_0}} \times \frac{f_s}{f_o} \tag{3.30}$$

Equation (3.30), plotted in Fig. 3.12, gives the output characteristics of the converter. It represents a hyperbola and for a given value of μ_o, the product $q\,\overline{I'_0}$ is constant.

Fig. 3.12 Output characteristics

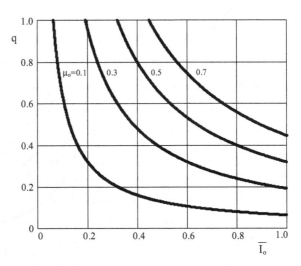

3.3.6 Normalized Output Power

The normalized output power is found multiplying Eq. (3.30) with the normalized output voltage.

$$\overline{P_o} = \frac{P_o \sqrt{L_r/C_r}}{V_1^2} = \frac{2}{\pi} \times \frac{f_s}{f_o} \tag{3.31}$$

3.3.7 Switches RMS Current

The switches RMS current is calculated by (3.32).

$$\overline{I_{S\,RMS}} = \frac{\overline{I'_{o\,RMS}}}{\sqrt{2}} = \frac{1}{\sqrt{2}} \times \sqrt{\frac{2}{T_s} \int_{t_0}^{t_1} \overline{i_{Lr}(t)}^2\, dt + \int_{t_1}^{t_2} \overline{i_{Lr}(t)}^2\, dt} \tag{3.32}$$

Substituting Eqs. (3.8) and (3.20) in Eq. (3.32) yields

$$\overline{I_{S\,RMS}} = \frac{1}{\sqrt{2}}$$

$$\times \sqrt{\frac{1}{\pi}\frac{f_s}{f_o}\left[\frac{(2-q)^2}{2}\left[\pi - \cos^{-1}\left(\frac{q}{2-q}\right)\right] + \left(q + \frac{8}{3\,q} - \frac{8}{3}\right)\sqrt{1-q}\right]} \tag{3.33}$$

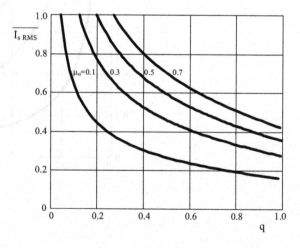

Fig. 3.13 Switches RMS current as a function of the static gain (q), taking μ_o as a parameter

Hence,

$$\overline{I_{S\,RMS}} = \frac{I_{S\,RMS}\,z}{V_1}$$

$$= \sqrt{\frac{1}{2\pi}\frac{f_s}{f_o}\left[\frac{(2-q)^2}{2}\left[\pi - \cos^{-1}\left(\frac{q}{2-q}\right)\right] + \left(q + \frac{8}{3q} - \frac{8}{3}\right)\sqrt{1-q}\right]}$$

(3.34)

The switches RMS current characteristics, as a function of the static gain, taking μ_o as a parameter, are plotted in Fig. 3.13.

3.3.8   Clamping Diodes Average Current

The clamping diodes average current is determined by:

$$I_{DC} = \frac{1}{T_s}\int_{t_1}^{t_2} i_{Lr}(t)\,dt$$

(3.35)

Substituting (3.20) in (3.35) gives:

$$\overline{I_{DC}} = \frac{I_{DC}\,z}{V_1} = \frac{1}{\pi}\frac{(1-q)}{q}\frac{f_s}{f_o}$$

(3.36)

The clamping diodes average current as a function of the static gain (q), taking μ_o, as a parameter, is plotted in Fig. 3.14.

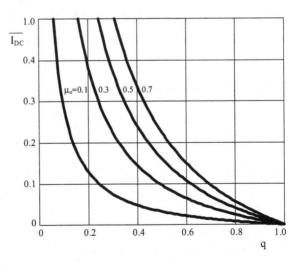

Fig. 3.14 Clamping diodes average current as a function of the static gain (q), taking μ_o as a parameter

3.4 CVC-SRC Topological Variations

Topological variations of the half bridge capacitor voltage-clamped series resonant converter are shown in Fig. 3.15. The operation is similar to the topology analyzed in previous subsections. Noted that these topologies do not need an input voltage with a middle point.

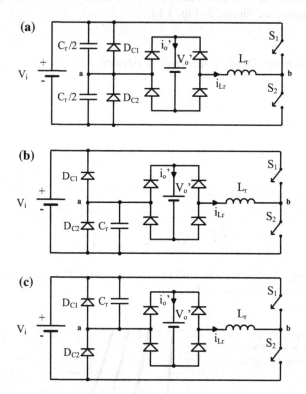

Fig. 3.15 Half bridge CVC-SRC topological variations

3.5 Design Example and Methodology

In this section, a design methodology and an example are presented, using the mathematical analysis results obtained in the Sect. 3.3. The converter is designed to operate in DCM according to the specifications given in Table 3.1.

A static gain (q) of 0.8 is chosen. The output DC voltage referred to the primary side of the transformer (V_o') is calculated as follows:

$$V_o' = q\, V_1 = 0.6 \times \frac{400}{2} = 160\ V$$

The transformer turns ratio (n) and the output current referred to the primary side of the transformer (I_o') are given by:

$$n = \frac{N_1}{N_2} = \frac{V_o'}{V_o} = \frac{160}{50} = 3.2$$

$$I_o' = \frac{I_o}{n} = \frac{10}{3.2} = 3.125\ A$$

A normalized switching frequency ($\mu_{o\,max} = f_{s\,max}/f_o$) of 0.5 is chosen for the nominal power. Thus, the resonant frequency is:

$$f_o = \frac{f_{s\,max}}{\mu_{o\,max}} = \frac{100 \times 10^3}{0.5} = 200\ kHz$$

Since

$$\frac{1}{\sqrt{L_r\, C_r}} = 2\pi\, f_o = 2 \times \pi \times 200 \times 10^3$$

We have

$$L_r\, C_r = 6.3325 \times 10^{-13}.$$

The average output current referred to the primary side of the transformer gives a second relation between the inductance L_r and the capacitance C_r. From the output characteristics, if q = 0.8 we have

Table 3.1 Design specifications

Input DC voltage (V_i)	400 V
Output DC voltage (V_o)	50 V
Nominal average output current (I_o)	10 A
Nominal power (P_o)	500 W
Maximum switching frequency ($f_{s\,max}$)	100 kHz
Minimum switching frequency ($f_{s\,min}$)	20 kHz

$$\overline{I}_o = \frac{I'_o \sqrt{L_r/C_r}}{V_1} = 0.4$$

Thus,

$$I'_o = \frac{I_o}{N_1/N_2} = \frac{10}{3.2} = 3.125 \text{ A}$$

And

$$\sqrt{\frac{L_r}{C_r}} = 25.6$$

Finishing the calculations, we find

$$C_r = 31.085 \text{ nF}$$

$$L_r = 20.372 \text{ μH}$$

The switches S_1 and S_2 conduction time (Δt_1) as well as the clamping diodes D_{C1} and D_{C2} conduction time (Δt_2) are calculated as follows:

$$\Delta t_1 = \frac{1}{\omega_0}\left[\pi - \cos^{-1}\left(\frac{q}{2-q}\right) + \frac{2\sqrt{1-q}}{q}\right] = 2.72 \text{ μs}$$

$$\Delta t_2 = \frac{2\sqrt{1-q}/q}{\omega_0} = 0.8897 \text{ μs}$$

The initial conditions, the inductor peak current, the switches RMS current and diodes average current, for nominal power are calculated with the equations obtained in Sect. 3.3 and it results in:

$$I_{Lr} = 9.375 \text{ A} \quad I_1 = 6.98 \text{ A} \quad I_{S\,RMS} = 3.351 \text{ A} \quad I_{DC} = 0.311 \text{ A}$$

For the minimum switching frequency of 20 kHz, the normalized switching frequency ratio μ_o is:

$$\mu_{o\,min} = \frac{f_{s\,min}}{f_o} = 0.1$$

For this normalized frequency, the output power is:

$$P_{o\,min} = \frac{2}{\pi} \times \frac{f_{s\,min}}{f_o} \times \frac{V_1^2}{Z} = 100 \text{ W}$$

The initial conditions, the switches RMS current and diodes average current, for minimum power, are calculated with the equations obtained in Sect. 3.3, as follows:

$$I_{Lr} = 9.375\,A \quad I_1 = 6.98\,A \quad I_{S\,RMS} = 1.499\,A \quad I_{DC} = 0.062\,A$$

3.6 Simulation Results

The HB CVC-SRC presented in Fig. 3.16 is simulated to validate the analysis and design example presented in Sect. 3.5 in the nominal and minimum power.

Figure 3.17 shows the resonant capacitor voltage, the resonant inductor current, the AC voltage v_{ab} and the output current i'_o in the nominal power. It may be noticed that the converter is operating in DCM, as predicted. Figure 3.18 shows the voltage and current in switches S_1 and S_2 and clamping diodes D_{C1} and D_{C2}. ZCS commutation is achieved when it turns on and off.

Table 3.2 presents theoretical and simulated parameters and component stresses at nominal power to validate the mathematical analysis.

Figure 3.19 shows the resonant capacitor voltage, the resonant inductor current, the AC voltage v_{ab} and the output current i'_o, at minimum power. Figure 3.20 shows the voltage and current in switches S_1 and S_2 and clamping diodes D_{C1} and D_{C2}. ZCS commutation is also achieved when it turns on and off and the converter is operating in DCM. The inductor current peak value is the same at minimum and nominal power.

Table 3.3 presents theoretical and simulated parameters and component stresses with minimum power to validate the mathematical analysis.

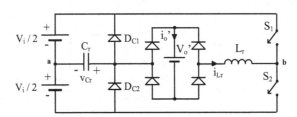

Fig. 3.16 Simulated circuit

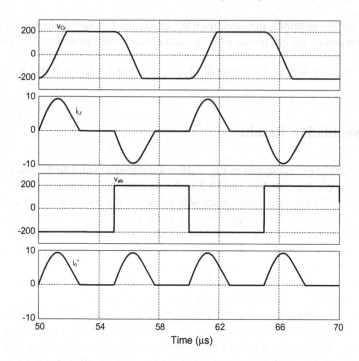

Fig. 3.17 Resonant capacitor voltage, resonant inductor current, AC voltage v_{ab} and output current i'_o, at nominal power

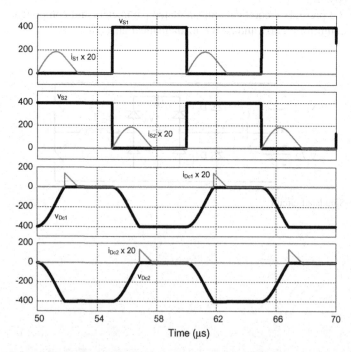

Fig. 3.18 Switches and clamping diodes soft commutation at nominal power: switches S_1 and S_2 voltage and current, clamping diodes D_{C1} and D_{C2} voltage and current

Table 3.2 Theoretical and simulated results at nominal power

	Theoretical	Simulated
I_o' [A]	3.125	3.10
I_{Lr} [A]	9.375	9.37
I_1 [A]	6.98	6.95
$I_{S\ RMS}$ [A]	3.351	3.35
I_D [A]	0.311	0.30

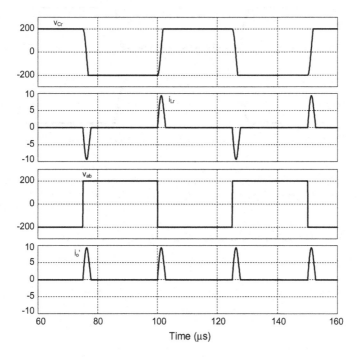

Fig. 3.19 Resonant capacitor voltage, resonant inductor current, AC voltage v_{ab} and output current i_o', at minimum power

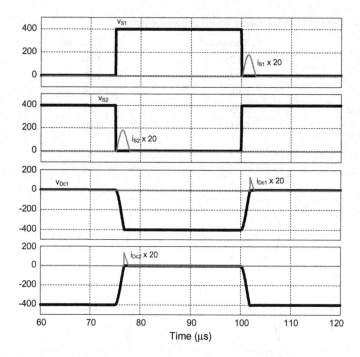

Fig. 3.20 Switches and clamping diodes soft commutation at the minimum power: switches S_1 and S_2 voltage and current, clamping diodes D_{C1} and D_{C2} voltage and current

Table 3.3 Theoretical and simulated results with minimum power

	Theoretical	Simulated
I'_o [A]	0.625	0.64
I_{Lr} [A]	9.375	9.37
I_1 [A]	6.98	6.95
$I_{S\ RMS}$ [A]	1.499	1.36
I_D [A]	0.062	0.053

3.7 Problems

(1) The half-bridge capacitor voltage clamped converter has the following specifications:

$$V_i = 400\ V \qquad V'_o = 150\ V \qquad C_r = 40\ nF$$
$$L_r = 30\ \mu H \quad f_s = 100 \times 10^3\ Hz$$

Calculate:

(a) The resonant frequency;
(b) The output average current and power;
(c) The resonant inductor peak current.
    Answers: (a) $f_o = 145.3$ kHz; (b) $I_o = 4.27$ A; $P_o = 640$ W; (c) $I_{Lr} = 9.13$ A.
(2) For the previous exercise, calculate the maximum switching frequency and the output power at this point of operation.
    Answers: (a) $f_{smax} = 128.7$ kHz; (b) $P_{omax} = 823.42$ W.
(3) Consider the topological variation of the HB CVC-SRC shown in Fig. 3.21. Show that this topology is similar to the one presented in this chapter and the same equations can be applied. Comment the advantages and disadvantages of this topology.
(4) For the topological variation presented in Fig. 3.22 and with the following parameters, calculate:

$$I = 5 \text{ A} \quad V_o' = 15 \text{ V} \quad C_r/2 = 20 \text{ nF} \quad C_i = 1 \text{ μF}$$

$L_r = 30$ μH and $f_s = 100 \times 10^3$ Hz, calculate

(a) The input voltage;
(b) The output power.
    Answers: (a) $V_i = 500$ V and (b) $P_o = 1000$ W.

Fig. 3.21 HB CVC-SRC topological variation

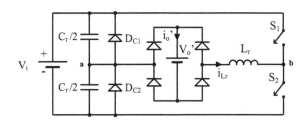

Fig. 3.22 HB CVC-SRC topological variation

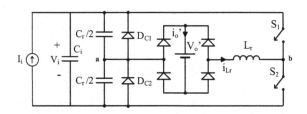

(5) A series-resonant converter with isolating transformer and voltage clamping of the resonant capacitor has the following parameters:

$$V_i = 60 \text{ V}, \quad V_o = 400 \text{ V}, \quad \mu_o = 0.6, \quad q = 0.8, \quad f_s = 120 \text{ kHz} \quad \text{and} \quad P_o = 300 \text{ W}.$$

Calculate: (a) L_r, (b) C_r; and (c) (N_p/N_s).
Answers: (a) $L_r = 0.512 \ \mu H$; (b) $C_r = 699.5 \text{ nF}$ and (c) $n = 0.06$.

Reference

1. Agrawal, J.P., Lee, C.Q.: Capacitor voltage clamped series resonant power supply with improved cross regulation. In: IEEE IAS Annual Meeting, pp. 1141–1146 (1989)

Chapter 4
Half Bridge CVC-PWM Series Resonant Converter

Nomenclature

V_i	DC Bus voltage
V_1	Half of the DC Bus voltage
V_o	Output DC voltage
P_o	Output power
$P_{o\ min}$	Minimum output power
C_o	Output filter capacitor
R_o	Output load resistor
q	Static gain
D	Duty cycle
D_{max}	Maximum duty cycle
D_{min}	Minimum duty cycle
f_s	Switching frequency
$f_{s\ max}$	Maximum switching frequency
$f_{s\ min}$	Minimum switching frequency
T_s	Switching period
f_o	Resonant frequency [Hz]
ω_o	Resonant frequency [rad/s]
μ_o	Frequency ratio
V_o'	Output DC voltage referred to the transformer primary side
i_o	Output current
i_o'	Output current referred to the transformer primary side
$i_o'\left(\overline{I_o'}\right)$	Average output current referred to the transformer primary and its normalized value
$I_{o\ min}$	Minimum average output current
$i_{o\ min}'$	S_2—main switches
S_3 and S_4	Auxiliary switches
v_{g1}, v_{g2}, v_{g3} and v_{g4}	Switches gate signals
S_1 and S_2	main switches
D_{C1} and D_{C2}	Clamping diodes
C_r	Resonant capacitors

© Springer International Publishing AG, part of Springer Nature 2019
I. Barbi and F. Pöttker, *Soft Commutation Isolated DC-DC Converters*,
Power Systems, https://doi.org/10.1007/978-3-319-96178-1_4

L_r	Resonant inductor (may include the transformer leakage inductance)
z	Characteristic impedance
i_{Lr}	Resonant inductor current
v_{Cr}	Resonant capacitor voltage
$I_1(\overline{I_1})$	Inductor current at the end of the first step of operation and its normalized value
$I_2(\overline{I_2})$	Resonant inductor peak current and its normalized value
$I_3(\overline{I_3})$	Inductor current at the end of the third step of operation and its normalized value
v_{ab}	AC voltage, between points "a" and "b"
v_{S1}, v_{S2}, v_{S3} and v_{S4}	Voltage across switches
i_{S1}, i_{S2}, i_{S3} and i_{S4}	Switches currents
Δt_1	Time interval of the first step of operation (t_1-t_0)
Δt_2	Time interval of the second step of operation (t_2-t_1)
Δt_3	Time interval of the third step of operation (t_3-t_2)
Δt_4	Time interval of the fourth step of operation (t_4-t_3)
Δt_5	Time interval of the fifth step of operation (t_5-t_4)
Δt_6	Time interval of the sixth step of operation (t_6-t_5)
Δt_7	Time interval of the seventh step of operation (t_7-t_6)
Δt_8	Time interval of the eighth step of operation (t_8-t_7)
Δt_9	Time interval of the ninth step of operation (t_9-t_8)
Δt_{10}	Time interval of the tenth step of operation $(t_{10}-t_9)$
$\phi, \theta_o, \theta_r, \theta$	State plane angle
$\overline{r_1}$ and $\overline{r_2}$	State plane radius

4.1 Introduction

The half bridge capacitor voltage-clamped series resonant converter studied in Chap. 3 had the frequency modulation as the main disadvantage. So, for a wide load range, the switching frequency has to vary as well. This means that the magnetic devices, such as resonant inductor and isolation transformer, are designed for a wide switching frequency range. This implies in a complex design of these components, in order to optimize losses, volume and size.

The half bridge capacitor voltage-clamped pulse width modulation series resonant converter [1], shown in Fig. 4.1, has two auxiliary switches, S_3 and S_4, that allow to control the power transfer to the load at constant switching frequency. To ensure soft commutation in the semiconductors turn on and off, the converter must operate in discontinuous conduction mode (DCM). The main switches (S_1 and S_2) commutate under zero current (ZCS) while the auxiliary switches (S_3 and S_4) commutate with zero voltage (ZVS).

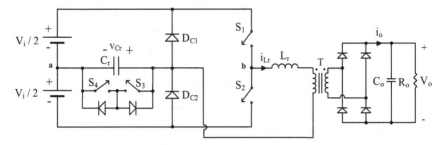

Fig. 4.1 Half bridge CVC-PWM series resonant converter

4.2 Circuit Operation

In this section, the topology shown in Fig. 4.2 is analyzed. In order to simplify, we assume that:

- all components are considered ideal;
- the converter is on steady-state operation;
- the output filter is represented as a DC voltage V_o', whose value is the output voltage referred to the primary winding of the transformer;
- the transformer magnetizing current is very small and can be neglected;
- current that flows through the switches is unidirectional, i.e. the switches allow the current to flow only in the direction of the arrow;
- the converter is controlled by pulse width modulation (PWM) and no dead time is considered at this point;
- the switches S_1 and S_2 operate with 50% duty cycle and the auxiliary switches S_3 and S_4 operate with variable duty-cycle, in order to control the power transfer to the load;

The converter is analyzed in DCM. In this conduction mode, the semiconductors' turn on and off are non-dissipative, switches S_1 and S_2 have zero current (ZCS) and switches S_3 and S_4 have zero voltage (ZVS).

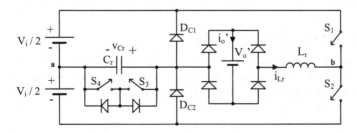

Fig. 4.2 Ideal half-bridge CVC-PWM series resonant converter with all parameters referred to the transformer primary winding

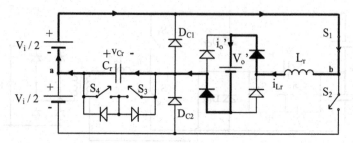

Fig. 4.3 Topological state for time interval Δt_1

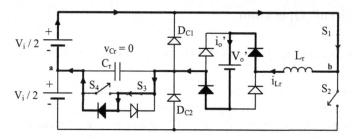

Fig. 4.4 Topological state for time interval Δt_2

(A) **Time Interval Δt_1 ($t_0 \leq t \leq t_1$)**

Time interval Δt_1, which converter topological state is shown in Fig. 4.3, starts at $t = t_0$, when switch S_1 is gated on. The resonant capacitor initial condition is $-V_i/2$ and the resonant inductor initial condition is zero. The circuit state variables evolve in a resonant way and this time interval ends in the instant $t = t_1$, when the capacitor voltage reaches zero.

(B) **Time Interval Δt_2 ($t_1 \leq t \leq t_2$)**

In the instant $t = t_1$, the resonant capacitor reaches zero, switch S_3 turns on with zero voltage (ZVS) and the inductor current starts to evolve linearly. This time interval controls the power transferred to the load. The topological state is shown in Fig. 4.4.

(C) **Time Interval Δt_3 ($t_2 \leq t \leq t_3$)**

At the instant $t = t_2$ switch S_3 is turned off with zero voltage (ZVS). During this time interval, shown in Fig. 4.5, the capacitor voltage and inductor current evolve in a resonant way. This time interval ends at the instant $t = t_3$, when the resonant capacitor reaches $V_i/2$.

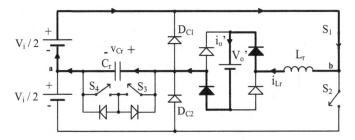

Fig. 4.5 Topological state for time interval Δt_3

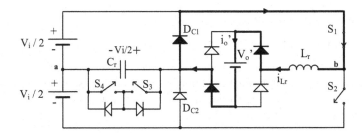

Fig. 4.6 Topological state for time interval Δt_4

(D) Time Interval Δt_4 ($t_3 \leq t \leq t_4$)

In the moment that the capacitor voltage reaches $V_i/2$, the clamping diode D_{C1} starts conducting, as shown in Fig. 4.6. The capacitor voltage remains clamped at $V_i/2$ and the inductor current decreases linearly. This time interval ends when the inductor current reaches zero, at the instant $t = t_4$.

(E) Time Interval Δt_5 ($t_4 \leq t \leq t_5$)

The topological state for time interval Δt_5 is shown in Fig. 4.7. At $t = t_4$, the inductor current reaches zero turning off the clamping diode D_{C1}. As the switch S_2

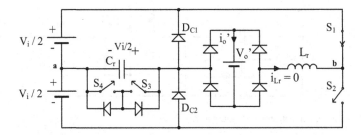

Fig. 4.7 Topological state for time interval Δt_5

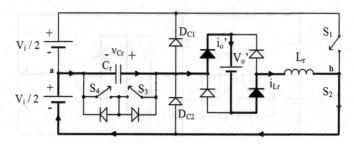

Fig. 4.8 Topological state for time interval Δt_6

is not gated on yet, the current does not flow through the circuit and the capacitor voltage remains clamped at $V_i/2$.

(F) **Time Interval Δt_6 ($t_5 \le t \le t_6$)**

Time interval Δt_6, shown in Fig. 4.8, starts at $t = t_5 = T_s/2$ when switch S_2 is turned on. The resonant capacitor initial condition is $+V_i/2$ and the resonant inductor initial condition is zero. The circuit evolves in a resonant way and this time interval ends when the capacitor voltage reaches zero.

(G) **Time Interval Δt_7 ($t_6 \le t \le t_7$)**

At $t = t_6$, the resonant capacitor reaches zero, switch S_4 turns on with zero voltage (ZVS) and the inductor current evolves linearly. This time interval also controls the power transferred to the load. The topological state for this time interval is presented in Fig. 4.9.

(H) **Time Interval Δt_8 ($t_7 \le t \le t_8$)**

At $t = t_7$, switch S_3 is turned off with zero voltage (ZVS). During this time interval, shown in Fig. 4.10, the capacitor voltage and inductor current evolve in a resonant way. This time interval ends when the resonant capacitor reaches $-V_i/2$, in the instant $t = t_8$.

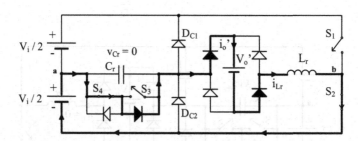

Fig. 4.9 Topological state for time interval Δt_7

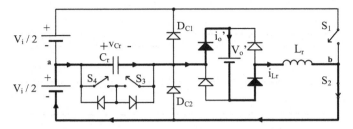

Fig. 4.10 Topological state for time interval Δt_8

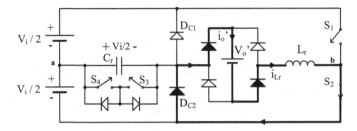

Fig. 4.11 Topological state for time interval Δt_9

(I) Time Interval Δt_9 ($t_8 \leq t \leq t_9$)

When the capacitor voltage reaches $-V_i/2$, at the instant $t = t_8$, the clamping diode D_{C2} is turned on, as shown in Fig. 4.11. The capacitor voltage is clamped at $-V_i/2$ and the inductor current decreases linearly. This time interval ends at the instant $t = t_9$, when the inductor current reaches zero.

(J) Time Interval Δt_{10} ($t_9 \leq t \leq t_{10}$)

The topological state for time interval Δt_{10} is shown in Fig. 4.12. It starts at $t = t_9$, when the inductor current reaches zero, turning off the clamping diode D_{C2}. As switch S_1 is not gated on yet, the current does not flow through the circuit and the capacitor voltage remains clamped at $-V_i/2$. The main waveforms along with time diagram are shown in Fig. 4.13.

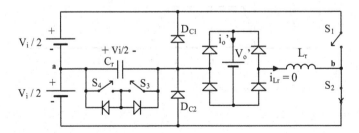

Fig. 4.12 Topological state for time interval Δt_{10}

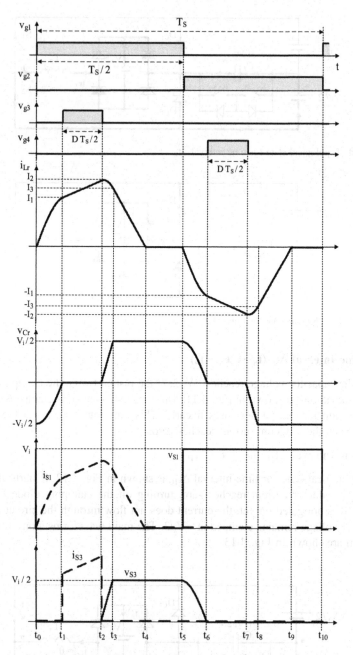

Fig. 4.13 Main waveforms and time diagram

4.3 Mathematical Analysis

In this section, using mathematical analysis, it will be obtained the resonant capacitor voltage and the resonant inductor current expressions, as well as the DCM constraints and the output characteristics. Since symmetry is assumed, only half of the switching cycle will be analyzed.

4.3.1 Time Interval Δt_1

In this time interval, the initial resonant inductor current and resonant capacitor voltage are: $\begin{cases} i_{Lr}(t_0) = 0 \\ v_{Cr}(t_0) = -V_1 \end{cases}$

From the equivalent circuit shown in Fig. 4.3, we have:

$$V_1 = L_r \frac{di_{Lr}(t)}{dt} + V_o' + v_{Cr}(t) \tag{4.1}$$

$$i_{Lr}(t) = C_r \frac{dv_{Cr}(t)}{dt} \tag{4.2}$$

Application of the Laplace Transform gives:

$$\frac{V_1 - V_o'}{s} = s\, L_r\, I_{Lr}(s) + V_{Cr}(s) \tag{4.3}$$

and

$$I_{Lr}(s) = s\, C_r\, V_{Cr}(s) + C_r\, V_1 \tag{4.4}$$

Applying the inverse Laplace Transform in Eqs. (4.3) and (4.4) and rearranging terms, we find:

$$\overline{v_{Cr}(t)} = \frac{v_{Cr}(t)}{V_1} = 1 - q - (2 - q)\cos(\omega_o t) \tag{4.5}$$

$$\overline{i_{Lr}(t)} = \frac{i_{Lr}(t)\, z}{V_1} = (2 - q)\, \text{sen}(\omega_o t) \tag{4.6}$$

where $V_1 = \frac{V_i}{2}$, $q = \frac{V_o'}{V_1}$, $z = \sqrt{\frac{L_r}{C_r}}$ and $\omega_o = \frac{1}{\sqrt{L_r \cdot C_r}}$.

This time interval ends when the capacitor voltage reaches zero. Thus, from Eq. (4.5) we have:

$$\omega_o \Delta t_1 = \cos^{-1}\left(\frac{1-q}{2-q}\right) \tag{4.7}$$

Substituting Eq. (4.7) into Eq. (4.6), it can be found the normalized resonant current in the end of this time interval, given by:

$$\overline{I_1} = \frac{I_1(t)\,z}{V_1} = \sqrt{3 - 2q} \tag{4.8}$$

4.3.2   Time Interval Δt_2

In this time interval, the initial resonant inductor current and resonant capacitor voltage are: $\begin{cases} i_{Lr}(t_1) = I_1 \\ v_{Cr}(t_1) = 0 \end{cases}$

From the equivalent circuit presented in Fig. 4.4, we have:

$$v_{Cr}(t) = 0 \tag{4.9}$$

$$i_{Lr}(t) = I_1 + \frac{V_1 - V_o'}{L_r}(t - t_1) \tag{4.10}$$

Normalizing (4.9) and (4.10) with V_1, we find:

$$\overline{v_{Cr}(t)} = \frac{v_{Cr}(t)}{V_1} = 0 \tag{4.11}$$

$$\overline{i_{Lr}(t)} = \frac{i_1(t)\,z}{V_1} = \overline{I_1} + (1 - q)\,\omega_o\,(t - t_1) \tag{4.12}$$

The duty-cycle D, that controls the power transferred to the load during this time interval, is given by:

$$D = \frac{\Delta t_2}{T_s/2} \tag{4.13}$$

Substituting Eq. (4.13) in Eq. (4.12) and rearranging terms, we find the inductor current in the end of this time interval, given by:

$$\bar{I_2} = \bar{I_1} + (1 - q) \frac{\pi D}{f_s/f_o} \tag{4.14}$$

4.3.3 Time Interval Δt_3

The initial conditions for time interval are: $\begin{cases} i_{Lr}(t_2) = I_2 \\ v_{Cr}(t_2) = 0 \end{cases}$

From the topological state of Fig. 4.5, we have:

$$V_1 = L_r \frac{di_{Lr}(t)}{dt} + V'_o + v_{Cr}(t) \tag{4.15}$$

$$i_{Lr}(t) = C_r \frac{dv_{Cr}(t)}{dt} \tag{4.16}$$

Applying the Laplace Transform yields:

$$\frac{V_1 - V'_o}{s} = s L_r I_{Lr}(s) - L_r I_2 + V_{Cr}(s) \tag{4.17}$$

$$I_{Lr}(s) = s C_r V_{Cr}(s) \tag{4.18}$$

Applying the inverse Laplace Transform in Eqs. (4.17) and (4.18), normalizing with V_1, and rearranging terms, we find:

$$\overline{v_{Cr}(t)} = \frac{v_{Cr}(t)}{V_1} = 1 - q - (1 - q)\cos(\omega_o t) + \bar{I_2}\,\text{sen}\,(\omega_o t) \tag{4.19}$$

$$\overline{i_{Lr}(t)} = \frac{i_{Lr}(t)\,z}{V_1} = (1 - q)\,\text{sen}\,(\omega_o t) + \bar{I_2}\cos(\omega_o t) \tag{4.20}$$

This time interval ends at $t = t_3$, when the capacitor voltage reaches V_1. Hence, from Eq. (4.19), we have:

$$\omega_o \Delta t_3 = \pi - \cos^{-1}\left(\frac{1 - q}{\sqrt{\bar{I_2}^2 + (1 - q)^2}}\right) - \cos^{-1}\left(\frac{q}{\sqrt{\bar{I_2}^2 + (1 - q)^2}}\right) \tag{4.21}$$

Substitution of Eq. (4.21) in Eq. (4.20) gives the normalized inductor current in the end of this time interval:

$$\overline{I_3} = \sqrt{\overline{I_2}^2 + (1-q)^2 - q^2} \tag{4.22}$$

4.3.4   Time Interval Δt_4

In this time interval, the initial resonant inductor current and resonant capacitor voltage are: $\begin{cases} i_{Lr}(t_3) = I_3 \\ v_{Cr}(t_3) = V_1 \end{cases}$

Based on the topological state shown in Fig. 4.6, we have:

$$v_{Cr}(t) = V_1 \tag{4.23}$$

$$i_{Lr}(t) = I_3 - \frac{V_o'}{L_r}(t - t_3) \tag{4.24}$$

Normalization with V_1, gives:

$$\overline{v_{Cr}(t)} = \frac{v_{Cr}(t)}{V_1} = 1 \tag{4.25}$$

$$\overline{i_{Lr}(t)} = \frac{i_{Lr}(t)\,z}{V_1} = \overline{I_3} - q\,\omega_o\,(t - t_3) \tag{4.26}$$

This time interval ends at the instant $t = t_3$, when the inductor current reaches zero. Thus, considering $\overline{i_{Lr}(t)} = 0$ in Eq. (4.26), we find:

$$\omega_o \Delta t_4 = \frac{\overline{I_3}}{q} \tag{4.27}$$

4.3.5   Normalized State-Plane Trajectory

The normalized state-plane trajectory for the Half Bridge CVC-PWM SRC is illustrated in Fig. 4.14.

Fig. 4.14 State-plane diagram of the half bridge CVC-PWM SRC

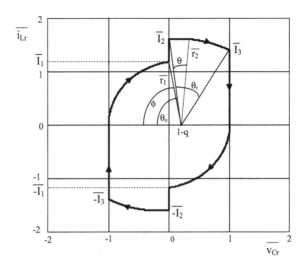

4.3.6 DCM Constraint

The switching period constraint at the boundary between discontinuous and continuous conductions mode is:

$$\frac{T_s}{2} = \Delta t_1 + \Delta t_2 + \Delta t_3 + \Delta t_4 \tag{4.28}$$

Assuming that the resonant frequency is much higher than the switching frequency, which implies in a very low value of the normalized switching frequency μ_o. Thus, the time interval $\Delta t_3 \cong 0$ can be neglected, to simplify.

Multiplying the terms of Eq. (4.28) by ω_o gives

$$\frac{\omega_o T_s}{2} = \omega_o (\Delta t_1 + \Delta t_2 + \Delta t_3 + \Delta t_4) \tag{4.29}$$

Substitution of Eqs. (4.7), (4.13) and (4.27) into Eq. (4.28), and doing appropriate algebraic manipulation gives:

$$\frac{\pi}{\mu_o} = \cos^{-1}\left(\frac{1-q}{2-q}\right) + \frac{\pi}{\mu_o} D_{max} + \sqrt{3 - 2q} + (1 - 2q)\frac{\pi}{\mu_o} D_{max} \tag{4.30}$$

Solving Eq. (4.30) for D_{max}, we obtain

$$D_{max} = \frac{\mu_o \cos^{-1}\left(\frac{1-q}{2-q}\right) + \mu_o \sqrt{3 - 2q} - \pi}{\pi(q - 2)} \tag{4.31}$$

Fig. 4.15 Maximum duty-cycle as a function of the static gain q, taking μ_o as a parameter

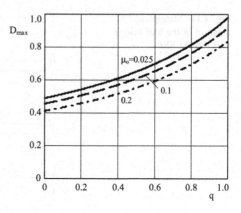

The maximum duty-cycle, as a function of the static gain, taking μ_o as a parameter given by Eq. (4.31), is plotted in Fig. 4.15. If this constraint is not respected, the converter will not operate in discontinuous conduction mode (DCM) as we assumed in the beginning of the theoretical analysis.

4.3.7 Output Characteristics

The normalized output current referred to the transformer primary side is shown in Fig. 4.16. Note that the period of this current waveform is half of the converter switching period.

The average value of the normalized output current $\overline{I'_o}$ is determined by Eq. (4.32).

$$\overline{I'_o} = \frac{2}{T} \left(\int_{t_0}^{t_1} \overline{i_{Lr}(t)}\, dt + \int_{t_1}^{t_2} \overline{i_{Lr}(t)}\, dt + \int_{t_2}^{t_3} \overline{i_{Lr}(t)}\, dt + \int_{t_3}^{t_4} \overline{i_{Lr}(t)}\, dt \right) \qquad (4.32)$$

Substituting (4.6), (4.12), (4.20) and (4.26) in (4.32) and rearranging terms, yields:

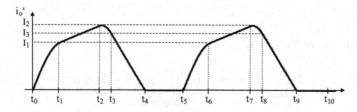

Fig. 4.16 Normalized output current referred to the transformer primary side

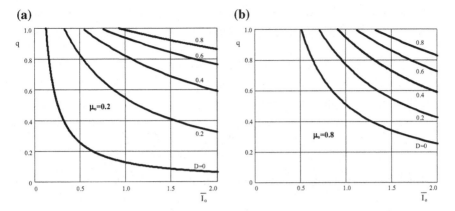

Fig. 4.17 Output characteristics taking the duty-cycle as parameter for **a** $\mu_o = 0.2$ and **b** $\mu_o = 0.8$

$$\overline{I}_o' = \frac{I_o' z}{V_1} = \frac{2}{\pi}\frac{\mu_o}{q} + \frac{(1-q)}{2q}\frac{\pi D^2}{\mu_o} + \frac{\sqrt{3-2q}}{q}D \qquad (4.33)$$

Equation (4.33) represents the normalized load current as a function of the static gain (q), taking the duty-cycle (D) and the normalized switching frequency (μ_o) as parameters. This means that this converter can be controlled by frequency modulation or pulse-width modulation. The output characteristics, given by Eq. (4.33) are plotted in Fig. 4.17.

4.4 Numerical Example

The converter shown in Fig. 4.1 operates with the specifications given by Table 4.1.

Table 4.1 Design specifications

Input DC voltage (V_i)	400 V
Output DC voltage (V_o)	50 V
Nominal average output current (I_o)	10 A
Nominal power (P_o)	500 W
Minimum power ($P_{o\ min}$)	100 W
Resonant frequency (f_o)	400 kHz
Static gain (q)	0.6
Transformer turns ratio (n)	3.2
Frequency ratio (μ_o)	0.1
Duty cycle (D)	0.4

Find: (a) the inductance of the resonant inductor L_r, the capacitance of the resonant capacitor C_r, and (b) the duty-cycle for operation at minimum power

Solution:

The output DC voltage referred to the primary side of the transformer (V'_o) is:

$$V'_o = q\,V_1 = 0.6 \times \frac{400}{2} = 160 \text{ V}$$

The normalized output current is:

$$\overline{I'_o} = \frac{I'_o\, z}{V_1} = \frac{2}{\pi} \times \frac{1}{q} \times \frac{f_s}{f_o} + \frac{(1-q)}{2q} \times \frac{\pi\,D^2}{f_s/f_o} + \frac{\sqrt{3-2q}}{q} \times D = 1.3$$

Thus, the average output current referred to the primary side of the transformer is:

$$I'_o = \frac{I_o}{N_1/N_2} = \frac{10}{3.2} = 3.125 \text{ A}$$

Hence,

$$z = \sqrt{\frac{L_r}{C_r}} = \frac{\overline{I'_o}\, V_1}{I'_o} = 83.2$$

Consequently,

$$L_r = 6922.24 C_r$$

We know that

$$L_r C_r = \frac{1}{(2\pi f_o)^2} = 1.58 \times 10^{-13}$$

Thus:

$$C_r = 4.782 \text{ nF}$$

$$L_r = 33.1 \text{ }\mu\text{H}$$

(b) The load current at the minimum power is

$$I_{o\,min} = \frac{100}{50} = 2 \text{ A}$$

Hence,

$$I'_{o\,min} = \frac{I_{o\,min}}{N_1/N_2} = \frac{2}{3.2} = 0.625 \text{ A}$$

and

$$\overline{I'_{o\,min}} = \frac{I'_{o\,min}\sqrt{L_r/C_r}}{V_1} = \frac{0.625 \times 83.2}{200} = 0.26$$

At minimum power we have:

$$\overline{I'_{o\,min}} = \frac{2}{\pi} \times \frac{1}{q} \times \frac{f_s}{f_o} + \frac{(1-q)}{2q} \times \frac{\pi D_{min}^2}{f_s/f_o} + \frac{\sqrt{3-2q}}{q} \times D_{min}$$

Solving for D_{min} we find:

$$D_{min} = 0.0236$$

4.5 Simulation Results

The Half Bridge CVC-PWM series resonant converter shown in Fig. 4.18 was simulated with the parameters of Sect. 4.4, to validate the analysis.

Figure 4.19 shows the resonant capacitor voltage, the resonant inductor current, the AC voltage v_{ab} and the output current I'_o, with nominal power. Note that the converter is in DCM. Figure 4.20 shows the voltage and current in switches S_1 and S_2.

Figure 4.21 shows the voltage and current in the auxiliary switch S_3 during the commutation under zero voltage.

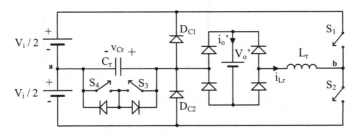

Fig. 4.18 Simulated circuit

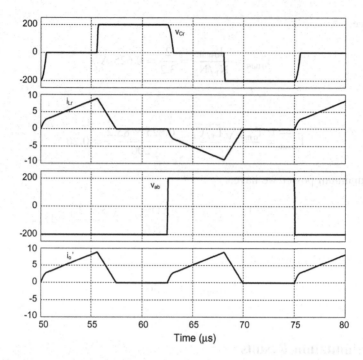

Fig. 4.19 Resonant capacitor voltage, resonant inductor current, AC voltage v_{ab} and output current i_o', with nominal power

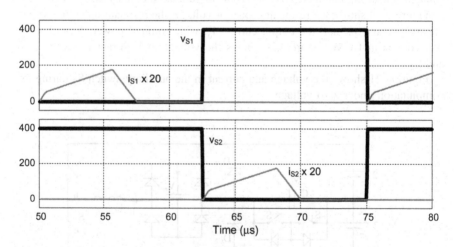

Fig. 4.20 Voltage and current in switches S_1 and S_2

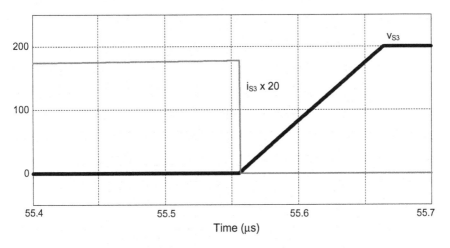

Fig. 4.21 A detail of the switch S_3 ZVS commutation turn off

4.6 Problems

(1) A HB CVC-PWM series resonant converter and its switches gate signals are shown in Fig. 4.22, with the following parameters:

$$V_i = 400 \text{ V} \quad V'_o = 140 \text{ V} \quad C_r = 30 \text{ nF}$$
$$L_r = 10 \text{ μH} \quad f_s = 60 \text{ kHz} \quad DT_s/2 = 4.63 \text{ μs}$$

Find:

(a) The average output current;
(b) The output power.

Answers: (a) $I_o = 19.123$ A; (b) $P_o = 2.68$ kW

(2) The HB CVC-PWM series resonant converter shown in Fig. 4.23 along with its gate drive signals has the specifications bellow:

$$V_i = 400 \text{ V} \quad R_o = 10 \text{ Ω} \quad C_o = 30 \text{ μF} \quad C_r = 30 \text{ nF}$$
$$L_r = 20 \text{ μH} \quad f_s = 40 \text{ kHz} \quad D = 0.44$$

Find:

(a) The output voltage referred to the transformer primary side;
(b) The output power;
(c) The duty cycle.

Answers: (a) $V'_o = 125$ V; (b) $P_o = 1563$ W; (c) $D = 0.364$.

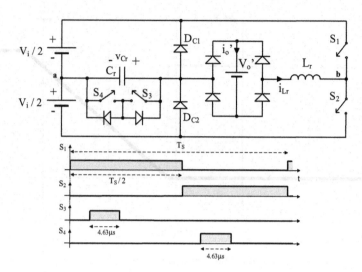

Fig. 4.22 HB CVC-PWM SRC and its drive signals

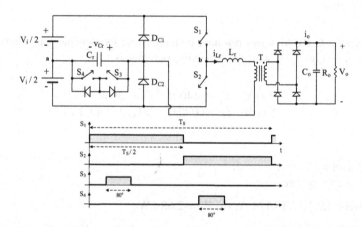

Fig. 4.23 HB CVC-PWM SRC and its drive signals

(3) Consider the topological variation of the HB CVC-SRC presented in Fig. 4.24. Describe the time intervals and the main waveforms for one switching period. Show that this topology has the same static gain expression like the converter presented in this chapter.

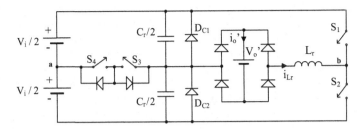

Fig. 4.24 HB CVC-PWM SRC topological variation

Reference

1. Freitas Vieira, J.L., Barbi, I.: Constant frequency PWM capacitor voltage-clamped series resonant power supply. IEEE Trans. Power Electron. **8**(2), 120–126 (1993)

Chapter 5
Series Resonant Converter Operating Above the Resonant Frequency

Nomenclature

V_i	Input DC voltage
V_o	Output DC voltage
P_o	Output power
C_o	Output filter capacitor
R_o	Output load resistor
q and q_{Co}	Static gain
f_s	Switching frequency [Hz]
ω_s	Switching frequency [rad/s]
$f_{s\ max}$	Maximum switching frequency
$f_{s\ min}$	Minimum switching frequency
T_s	Switching period
f_o	Resonant frequency [Hz]
ω_o	Resonant frequency [rad/s]
μ_o	Frequency ratio
z	Characteristic impedance
t_d	Dead time
T	Transformer
n	Transformer turns ratio
N_1 and N_2	Transformer windings
V'_o	Output DC voltage referred to the transformer primary side
i_o	Output current
i'_o	Output current referred to the transformer primary side
$I'_o\ \left(\overline{I'_o}\right)$	Average output current referred to the transformer primary and its normalized value
S_1, S_2, S_3 and S_4	Switches
D_1, D_2, D_3 and D_4	Diodes
C_1, C_2, C_3 and C_4	Capacitors
C_r	Resonant capacitor
v_{Cr}	Resonant capacitor voltage

© Springer International Publishing AG, part of Springer Nature 2019
I. Barbi and F. Pöttker, *Soft Commutation Isolated DC-DC Converters*,
Power Systems, https://doi.org/10.1007/978-3-319-96178-1_5

V_{Co}	Resonant capacitor peak voltage
V_{C1}	Resonant capacitor voltage at the end of time interval 1 and three
Q	Resonant capacitor charge
L_r	Resonant inductor (may include the transformer leakage inductance)
i_{Lr}	Resonant inductor current
I_{Lr}	Inductor fundamental peak current
$I_1 \left(\overline{I_1} \right)$	Inductor current at the end of the first and fourth step of operation and its normalized value
v_{ab}	AC voltage, between points "a" and "b"
v_{cb}	AC voltage, between points "c" and "b"
v_{ab1}	Fundamental AC voltage, between points "a" and "b"
v_{cb1}	Fundamental AC voltage, between points "c" and "b"
v_{S1} and v_{S2}	Voltage across switches S_1 and S_2
v_{g1} and v_{g2}	Switches drive signals
i_{S1} and i_{S2}	Current in the switches S_1 and S_2
i_{C1} and i_{C2}	Capacitors current
x_{Lr}, x_{CR} and x	Reactance
R_1, R_2	State plane radius
γ, β, θ	State plane angles
Δt_1	Time interval of the first step of operation in $(t_1–t_0)$
Δt_2	Time interval of the second step of operation $(t_2–t_1)$
Δt_3	Time interval of the third step of operation $(t_3–t_2)$
Δt_4	Time interval of the fourth step of operation $(t_4–t_3)$

5.1 Introduction

In Chap. 2, the series resonant converter was studied operating with a switching frequency bellow the resonant frequency, leading to zero current switching (ZCS). In this chapter, it will be analyzed the series resonant converter operating with a switching frequency above the resonant frequency. As a result, the switches commutate with zero voltage (ZVS) and the switches' intrinsic capacitance may be incorporated into the commutation process.

The power stage diagrams of the Full-Bridge ZVS and Half-Bridge ZVS series resonant converters operating above the resonant frequency are shown in Figs. 5.1 and 5.2, respectively.

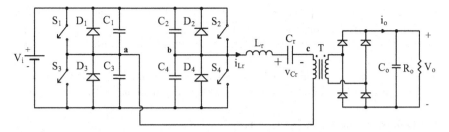

Fig. 5.1 Full-bridge ZVS series resonant converter

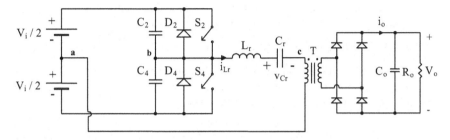

Fig. 5.2 Half-bridge ZVS series resonant converter

5.2 Circuit Operation

In this section, the converter shown in Fig. 5.3, without the commutation capacitor, is analyzed (the soft commutation analysis is presented in Sect. 5.4). The following assumptions are made:

- all components are considered ideal;
- the converter is on steady-state operation;

Fig. 5.3 Half-bridge series resonant converter

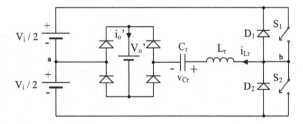

- the output filter is represented as a DC voltage V'_o, whose value is the output voltage referred to the primary of the transformer;
- current that flows through the switches is unidirectional, i.e. the switches allow the current to flow only in the direction of the arrow;
- the converter is controlled by frequency modulation and the switches operate at 50% duty cycle without dead time between the gate signals of the switches.

(A) Time Interval Δt_1 ($t_0 \leq t \leq t_1$)

The topological state for this time interval is presented in Fig. 5.4. It begins when the inductor current reaches zero at $t = t_0$. Switch S_1, that was already gated on, conducts the inductor current. The inductor current and capacitor voltage evolve in a resonant way. The DC bus delivers energy to the load.

(B) Time Interval Δt_2 ($t_1 \leq t \leq t_2$)

At $t = t_1$ switch S_1 is turned off and switch S_2 is gated on. As the inductor current is still positive, it starts to flow through diode D_2, as shown in Fig. 5.5. The resonant tank delivers energy to the DC bus and the load.

Fig. 5.4 Topological state for time interval Δt_1

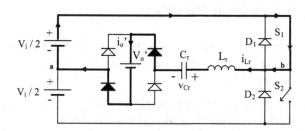

Fig. 5.5 Topological state for time interval Δt_2

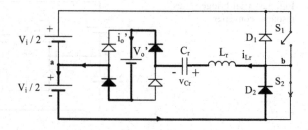

(C) **Time Interval Δt_3 ($t_2 \leq t \leq t_3$)**

Time interval Δt_3 begins when the inductor current reaches zero at $t = t_2$, as shown in Fig. 5.6. Switch S_2, that was already gated on, conducts the inductor current. The inductor current and capacitor voltage evolve in a resonant way. The DC bus delivers energy to the load.

(D) **Time Interval Δt_4 ($t_3 \leq t \leq t_4$)**

At $t = t_3$ switch S_2 is turned off and switch S_1 is gated on. As the inductor current is negative, it starts to flow through diode D_1, as shown in Fig. 5.7. The resonant tank delivers energy to the DC bus and to the load. The main waveforms and the time diagrams are shown in Fig. 5.8.

Fig. 5.6 Topological state for time interval Δt_3

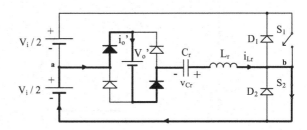

Fig. 5.7 Topological state for time interval Δt_4

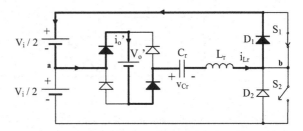

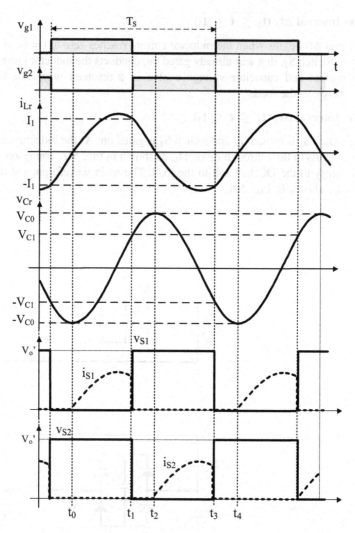

Fig. 5.8 Main waveforms and time diagram of the series-resonant converter operating with switching frequency above the resonant frequency

5.3   Mathematical Analysis

In this section, the resonant capacitor voltage and resonant inductor current equations are obtained as well as the output characteristics. Since the converter is symmetrical, only half of the switching cycle is analyzed.

(A) **Time Interval Δt_1**

In this time interval, the initial current in the resonant inductor and the initial voltage across the resonant capacitor are: $\begin{cases} i_{Lr}(t_0) = 0 \\ v_{Cr}(t_0) = -V_{C0} \end{cases}$

From the equivalent circuit shown in Fig. 5.4, we can write the differential equations:

$$\frac{V_i}{2} = L_r \frac{di_{Lr}(t)}{dt} + v_{Cr}(t) + V_o' \tag{5.1}$$

$$i_{Lr}(t) = C_r \frac{dv_{Cr}(t)}{dt} \tag{5.2}$$

The Laplace Transform of Eqs. (5.1) and (5.2) are:

$$\frac{V_1 - V_o'}{s} = s L_r i_{Lr}(s) + v_{Cr}(s) \tag{5.3}$$

$$i_{Lr}(s) = s C_r v_{Cr}(s) + C_r V_o' \tag{5.4}$$

where: $V_1 = V_i/2$.

Substituting (5.4) in (5.3) and applying the inverse Laplace Transform we obtain:

$$v_{Cr}(t) = V_1 - V_o' - \left(V_1 - V_o' + V_{C0}\right) \cos\left(\omega_0 t\right) \tag{5.5}$$

$$i_{Lr}(t)z = \left(V_1 - V_o' + V_{C0}\right) \text{sen}\left(\omega_0 t\right) \tag{5.6}$$

where $z = \sqrt{\frac{L_r}{C_r}}$ and $\omega_o = \frac{1}{\sqrt{L_r C_r}}$.

Normalizing (5.5) and (5.6) with V_1, we find:

$$\overline{v_{Cr}(t)} = \frac{v_{Cr}(t)}{V_1} = 1 - q - \left(1 - q + \overline{V_{C0}}\right) \cos(\omega_0 t) \tag{5.7}$$

$$\overline{i_{Lr}(t)} = \frac{i_{Lr}(t)z}{V_1} = \left(1 - q + \overline{V_{C0}}\right) \text{sen}(\omega_0 t) \tag{5.8}$$

where $q = \frac{V_o'}{V_1}$.

The time interval Δt_1 is:

$$\Delta t_1 = t_1 - t_0 = \frac{\theta}{\omega_o} \tag{5.9}$$

At the end of this interval the inductor current and capacitor voltage are:

$$\overline{I_1} = \left(1 - q + \overline{V_{C0}}\right) \mathrm{sen}(\theta) \qquad (5.10)$$

$$\overline{V_1} = (1 - q) - \left(1 - q + \overline{V_{C0}}\right) \cos(\theta) \qquad (5.11)$$

To obtain the state-plane trajectory for this time interval, we define:

$$\overline{v_{Cr}(t)} + j\overline{i_{Lr}(t)} = (1 - q) - \left(1 - q + \overline{V_{C0}}\right) e^{-jw_0 t} \qquad (5.12)$$

Equation (5.12) represents a circle, with the center at coordinates $(1 - q)$ and radius equal to $\left(1 - q + \overline{V_{C0}}\right)$.

(B) Time Interval Δt_2

The initial current in the resonant inductor and voltage across the resonant capacitor, in this time interval, are: $\begin{cases} i_{Lr}(t_2) = I_1 \\ v_{Cr}(t_2) = V_{C1} \end{cases}$

From the equivalent circuit shown in Fig. 5.5, we have:

$$V_1 = -L_r \frac{di_{Lr}(t)}{dt} - v_{Cr}(t) - V_o' \qquad (5.13)$$

$$i_{Lr}(t) = C_r \frac{dv_{Cr}(t)}{dt} \qquad (5.14)$$

Applying the Laplace Transform, we find:

$$\frac{V_1 + V_o'}{s} = -s\, L_r\, i_{Lr}(s) + L_r\, I_1 - v_{Cr}(s) \qquad (5.15)$$

$$i_{Lr}(s) = s\, C_r\, v_{Cr}(s) - C_r\, V_{C1} \qquad (5.16)$$

Substituting (5.16) into (5.15) and applying the inverse Laplace Transform, we obtain:

$$v_{Cr}(t) = -V_1 - V_o' - \left(-V_1 - V_o' - V_{C1}\right) \cos(\omega_0 t) + I_1\, z\, \mathrm{sen}\,(\omega_0 t) \qquad (5.17)$$

$$i_{Lr}(t)z = -\left(V_1 + V_o' + V_{C1}\right) \mathrm{sen}\,(\omega_0 t) + I_1\, z\, \cos\,(\omega_0 t) \qquad (5.18)$$

Normalizing with V_1 gives:

$$\overline{v_{Cr}(t)} = \frac{v_{Cr}(t)}{V_1} = -1 - q + \left(1 + q + \overline{V_{C1}}\right) \cos(\omega_0 t) + \overline{I_1}\, \mathrm{sen}\,(\omega_0 t) \qquad (5.19)$$

$$\overline{i_{Lr}(t)} = \frac{i_{Lr}(t)z}{V_1} = -\left(1 + q + \overline{V_{C1}}\right) \mathrm{sen}(\omega_0 t) + \overline{I_1} \cos(\omega_0 t) \qquad (5.20)$$

Time interval Δt_2 is:

$$\Delta t_2 = t_2 - t_1 = \frac{\gamma}{\omega_o} \tag{5.21}$$

The capacitor voltage in the end of this time interval is:

$$\overline{V_{C0}} = -1 - q + \left(1 + q + \overline{V_{C1}}\right)\cos(\gamma) + \overline{I_1}\,\text{sen}(\gamma) \tag{5.22}$$

To obtain the state-plane trajectory for this time interval, we define:

$$\overline{v_{Cr}(t)} + j\,\overline{i_{Lr}(t)} = -(1 + q) + \left[\left(1 + q + \overline{V_{C1}}\right) + j\overline{I_1}\right]e^{-j\omega_o t} \tag{5.23}$$

Equation (5.23) represents a circle, with the center at coordinates $-(1 - q)$ and radius equal to $\sqrt{\left(1 + q + \overline{V_{C1}}\right)^2 + \overline{I_1}^2}$.

(C) State Plane Trajectory for one Switching Cycle

The state plane trajectory for one switching cycle is shown in Fig. 5.9.

(D) Output Characteristics

For the normalized state plane shown in Fig. 5.9, the radius R_1 and R_2, and the angles θ and γ are given by Eqs. (5.24), (5.25), (5.26) and (5.27), respectively.

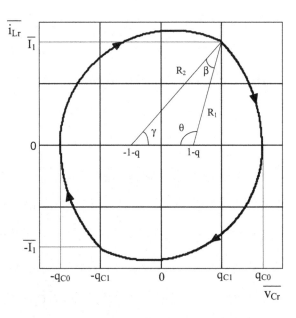

Fig. 5.9 State plane trajectory for the series resonant converter operating on steady-state, for $\mu_o \geq 1$

$$R_1 = q_{CO} + 1 - q \tag{5.24}$$

$$R_2 = q_{CO} + 1 + q \tag{5.25}$$

$$\theta = \omega_o \Delta t_1 \tag{5.26}$$

$$\gamma = \omega_o \Delta t_2 \tag{5.27}$$

In one switching period the following equation may be written:

$$\frac{1}{T_s} = \frac{1}{2(\Delta t_1 + \Delta t_2)} \tag{5.28}$$

Hence,

$$\mu_o = \frac{f_s}{f_o} = \frac{2\pi}{2(\Delta t_1 + \Delta t_2)} \times \frac{1}{\omega_o} \tag{5.29}$$

Therefore,

$$\frac{\mu_o}{\pi} = \frac{1}{\omega_o(\Delta t_1 + \Delta t_2)} = \frac{1}{\theta + \gamma} \tag{5.30}$$

In half of the switching period the capacitor charge (Q) is:

$$Q = 2 C_r V_{Co} = \frac{2}{T_s} \int_0^{T_s/2} i_{Lr}(t)dt \tag{5.31}$$

The average load current is determined by:

$$I_o' = \frac{2}{T_s} \int_0^{T_s/2} i_{Lr}(t)dt \tag{5.32}$$

Substituting (5.32) in (5.31) yields:

$$Q = 2 C_r V_{Co} = \frac{0}{2} I_o' \tag{5.33}$$

Thus,

$$V_{Co} = \frac{I_o'}{4 f_s C_r} \tag{5.34}$$

Normalizing (5.34) with V_1, we have:

$$q_{Co} = \frac{V_{Co}}{V_1} = \frac{I'_o}{4\,f_s\,C_r\,V_1} \tag{5.35}$$

The normalized average output current is:

$$\overline{I'_o} = \frac{z\,I'_o}{V_1} \tag{5.36}$$

Substituting (5.35) in (5.34), it can be found:

$$q_{Co} = \frac{V_{Co}}{V_1} = \frac{\overline{I'_o}\,V_1}{z} \times \frac{1}{4\,f_s\,C_r\,V_1} = \frac{\overline{I'_o}}{2} \times \frac{\pi}{2\pi\,f_s\,C_r\,z} = \frac{\overline{I'_o}}{2} \times \frac{\pi}{\mu_o} \tag{5.37}$$

Substitution of (5.36) in (5.24) and (5.25) yields:

$$R_1 = q_{co} + 1 - q \tag{5.38}$$

$$R_2 = q_{co} + 1 + q \tag{5.39}$$

From the state-plane trajectory in Fig. 5.9, using the law of cosines, we find:

$$2^2 = R_1^2 + R_2^2 - 2\,R_1 R_2 \cos\beta \tag{5.40}$$

Since the sum of the interior angles of a triangle is π, we have: $\beta = \pi - \gamma - \theta$ or $\beta = \pi - \frac{\pi}{\mu_o}$.

Let $\rho = \frac{\pi}{\mu_o}$. Thus, $\beta = \pi - \rho$.

With appropriate algebraic manipulation and rearranging terms, we find

$$R_1 = \frac{\overline{I'_o}}{2}\rho + 1 - q \tag{5.41}$$

$$R_2 = \frac{\overline{I'_o}}{2}\rho + 1 + q \tag{5.42}$$

$$\cos(\beta) = -\cos(\rho) \tag{5.43}$$

Substituting (5.41), (5.42) and (5.43) in (5.40) we obtain:

$$4 = \left(\frac{\overline{I'_o}}{2} \times \rho + 1 - q\right)^2 + \left(\frac{\overline{I'_o}}{2} \times \rho + 1 + q\right)^2 \\ + 2\left(\frac{\overline{I'_o}}{2} \times \rho + 1 - q\right)\left(\frac{\overline{I'_o}}{2} \times \rho + 1 + q\right)\cos(\rho) \tag{5.44}$$

Thus,

$$2\left(\frac{\overline{I'_o}\,\rho}{2}+1\right)^2 + 2\,q^2 + 2\left(\frac{\overline{I'_o}\,\rho}{2}+1-q\right) \times \left(\frac{\overline{I'_o}\,\rho}{2}+1+q\right)\cos(\rho) - 4 = 0$$

(5.45)

Rearranging terms, gives:

$$2\left(\frac{\overline{I'_o}\,\rho}{2}+1\right)^2 + 2\,q^2 + 2\left(\frac{\overline{I'_o}\,\rho}{2}+1\right)^2\cos(\rho) - 2q^2\,\cos(\rho) - 4 = 0$$

(5.46)

Letting $\frac{1-\cos(\rho)}{2} = \sin^2\left(\frac{\rho}{2}\right)$ and $\frac{1+\cos(\rho)}{2} = \cos^2\left(\frac{\rho}{2}\right)$ in Eq. (5.46) gives:

$$q^2\sin^2\left(\frac{\rho}{2}\right) = 1 - \left(\frac{\overline{I'_o}\,\rho}{2}+1\right)^2\cos^2\left(\frac{\rho}{2}\right)$$

(5.47)

Solving the algebraic Eq. (5.47) in relation to the static gain q, yields:

$$q = \sqrt{\frac{1 - \left(\frac{\overline{I'_o}\,\rho}{2}-1\right)^2\cos^2\left(\frac{\rho}{2}\right)}{\sin^2\left(\frac{\rho}{2}\right)}}$$

(5.48)

Fig. 5.10 Output characteristics of the ideal series resonant converter for $\mu_o \geq 1$

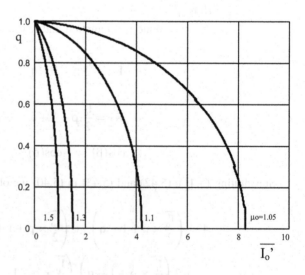

Substitution of $\rho = \pi/\mu_o$ in Eq. (5.48) gives:

$$q = \sqrt{\dfrac{1 - \left(\dfrac{\overline{I}_o \pi}{2\mu_o} + 1\right)^2 \cos^2\left(\dfrac{\pi}{2\mu_o}\right)}{\sin^2\left(\dfrac{\pi}{2\mu_o}\right)}} \qquad (5.49)$$

Equation (5.49) represents the static gain of the series resonant converter, with switching frequency larger than the resonant frequency, as a function of the normalized output current and having the normalized switching frequency μ_o as parameter. The output characteristics are plotted in Fig. 5.10. Note that the series resonant converter operating above the resonant frequency has a current source characteristic that can be beneficial for protection of the converter against overload or overcurrent, similar to the series resonant converter operating with switching frequency lower than the resonant frequency.

(E) **Analysis of the Series Resonant Converter for $\mu_o \geq 1$, Based in the First Harmonic Approximation**

In this section, the analysis of the series resonant operating above the resonant frequency, using the first harmonic approximation, is studied. The series resonant converter equivalent circuit is shown in Fig. 5.11.

The transformer is considered ideal and its magnetizing current is neglected. The amplitude of the fundamental components of the rectangular voltage v_{ab} and v_{cb} are given by expressions (5.50) and (5.51), respectively.

$$v_{ab1} = \frac{4}{\pi} V_1 \qquad (5.50)$$

$$v_{cb1} = \frac{4}{\pi} V'_o \qquad (5.51)$$

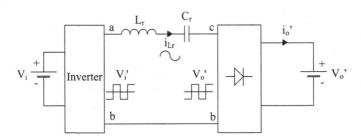

Fig. 5.11 Series resonant converter equivalent circuit

Defining the absolute values of x_{Lr} and x_{Cr} by the expressions (5.52) and (5.53), respectively.

$$|x_{Lr}| = L_r \omega_s = 2\pi f_s L_r \tag{5.52}$$

$$|x_{Cr}| = \frac{1}{C_r \omega_s} = \frac{1}{2\pi f_s C_r} \tag{5.53}$$

Hence, the absolute value of the equivalent reactance is given by expression (5.54).

$$|x| = |x_{Lr}| - |x_{cr}| \tag{5.54}$$

Therefore,

$$|x| = L_r \omega_s - \frac{1}{C_r \omega_s} \tag{5.55}$$

The output diode rectifier forces the current i_{Lr} to be in phase with the voltage v_{cb1}. Also, when the converter operates with $f_s \geq f_o$, the current i_{Lr} lags the voltage v_{ab1} by 90°.

Therefore, on steady state operation, we can represent the voltage and current by the phasor diagram shown in Fig. 5.12.

From the phasor diagram, we have:

$$v_{ab1}^2 = v_{cb1}^2 + (|x| I_{Lr})^2 \tag{5.56}$$

where I_{Lr} is the inductor's fundamental peak current.

Thus,

$$v_{cb1}^2 = v_{ab1}^2 - (|x| I_{Lr})^2 \tag{5.57}$$

Fig. 5.12 Phasor diagram for the first harmonic approximation

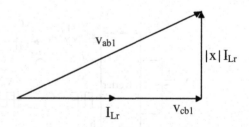

Substituting Eqs. (5.50) and (5.51) into (5.57), we obtain:

$$\left(\frac{4}{\pi}V_o'\right)^2 = \left(\frac{4}{\pi}V_1\right)^2 + (|x|I_{Lr})^2 \tag{5.58}$$

Thus,

$$\left(\frac{\frac{4}{\pi}V_o'}{\frac{4}{\pi}V_1}\right)^2 = 1 - \left(\frac{|x|I_{Lr}}{\frac{4}{\pi}V_1}\right)^2 \tag{5.59}$$

The static gain is defined by:

$$q = \frac{V_o'}{V_1} \tag{5.60}$$

Hence,

$$q^2 = 1 - \left(\frac{|x|I_{Lr}}{\frac{4}{\pi}V_1}\right)^2 \tag{5.61}$$

or yet,

$$q = \sqrt{1 - \left(\frac{|x|I_{Lr}}{\frac{4}{\pi}V_1}\right)^2} \tag{5.62}$$

Figure 5.13 shows the currents on the input and output of the diode rectifier.

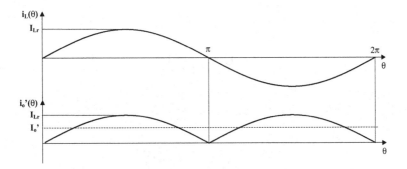

Fig. 5.13 Current on the input and output of the diode rectifier

The average value of the rectified current is:

$$I'_o = \frac{2}{\pi} I_{Lr}$$
(5.63)

Thus,

$$I_{Lr} = \frac{\pi}{2} I'_o$$
(5.64)

Substituting expression (5.64) in (5.62), we get:

$$q = \sqrt{1 - \left(\frac{|x| \pi I'_o}{2\frac{4}{\pi} V_1}\right)^2}$$
(5.65)

After appropriate simplifications, it can be found:

$$q = \sqrt{1 - \left(\frac{x I'_o \pi^2}{V_1 \; 8}\right)^2}$$
(5.66)

The expression (5.67) of the normalized current was already defined in previous sections:

$$\overline{I'_o} = \frac{z}{V_1} I'_o$$
(5.67)

Thus,

$$I'_o = \frac{V_1}{z} \overline{I'_o}$$
(5.68)

Or,

$$\left(\frac{|x| I'_o}{V_1}\right)^2 = \left(\frac{|x| \overline{I'_o}}{z}\right)^2$$
(5.69)

But,

$$\frac{|x|}{z} = \frac{1}{\sqrt{\frac{L_r}{C_r}}} \left(\omega_s L_r - \frac{1}{\omega_s C_r}\right)$$
(5.70)

With

$$\omega_o = \frac{1}{\sqrt{L_r C_r}}\tag{5.71}$$

and the appropriate algebraic manipulation of expressions (5.70) and (5.71), we find:

$$\left(\frac{|x|\,I'_o\,\pi^2}{V_1\,8}\right)^2 = \left(\frac{\pi^2}{8}\right)^2 \left(\frac{\omega_s}{\omega_o} - \frac{\omega_o}{\omega_s}\right)^2 \overline{I'}_o^2\tag{5.72}$$

Substituting Eq. (5.72) into (5.66), we find:

$$q = \sqrt{1 - \left[\frac{\pi^2}{8}\left(\frac{\omega_s}{\omega_o} - \frac{\omega_o}{\omega_s}\right)\overline{I'}_o\right]^2}\tag{5.73}$$

Since

$$\frac{\omega_o}{\omega_s} = \frac{f_o}{f_s}\tag{5.74}$$

We have:

$$q = \sqrt{1 - \frac{\pi^4}{64}\left(\frac{f_s}{f_o} - \frac{f_o}{f_s}\right)^2 \overline{I'}_o^2}\tag{5.75}$$

The normalized switching frequency is defined by:

$$\mu_o = \frac{f_s}{f_o}\tag{5.76}$$

Substitution of Eq. (5.76) in Eq. (5.75) yields

$$q = \sqrt{1 - \frac{\pi^4}{64}\left(\mu_o - \frac{1}{\mu_o}\right)^2 \overline{I'}_o^2}\tag{5.77}$$

Equation (5.77) represents the output characteristics of the series resonant converter, obtained for voltage and current fundamental components, for $f_s \geq f_o$. The output characteristics given by Eq. (5.77) are plotted in Fig. 5.14.

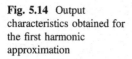

Fig. 5.14 Output
characteristics obtained for
the first harmonic
approximation

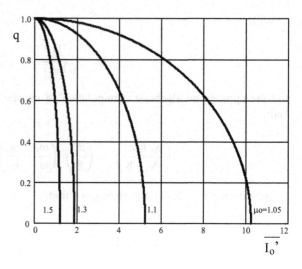

5.4 Commutation Analysis

Figure 5.15 shows the topological state and the relevant waveforms for the first commutation time interval that takes place between time interval Δt_1 and Δt_2 of Sect. 5.2. When switch S_1 is turned off but switch S_2 is not yet gated on (dead time). The inductor current during this time interval is I_1 and half of this current ($I_1/2$) flows through each of the commutation capacitors, charging C_1 from zero to V_i and discharging C_2 from V_i to zero. Figure 5.15 also shows the main waveforms where the soft commutation (ZVS) of the switches may be noted. As soon as capacitor C_2 is discharged, the diode D_2 starts to conduct the inductor current. The accomplishment of soft commutation requires that commutation capacitors' charge/discharge must finish before the dead time ends.

The second commutation time interval, whose topological state is shown in Fig. 5.16, takes place between time interval Δt_3 and Δt_4 of Sect. 5.2, when switch

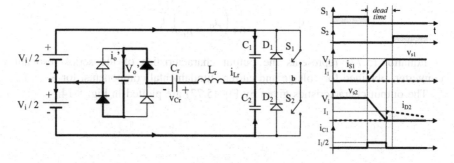

Fig. 5.15 Topological state and main waveforms for the first commutation time interval

S_2 is turned off but switch S_1 is not yet gated on (dead time). The value of the inductor current during this time interval is also I_1 and half of it flows through each one of the commutation capacitors, charging C_2 from zero to V_i and discharging C_1 from V_i to zero. Figure 5.16 also shows the relevant waveforms where the soft commutation (ZVS) of both switches may be noticed. At the instant the capacitor C_1 is fully discharged, diode D_1 starts conducting the inductor current. The capacitors must be fully charge/discharge before the dead time ends to realize soft commutation.

The commutation current I_1 may be found from the analysis of the state-plane trajectory in Fig. 5.9. According to the sine law we have:

$$\frac{2}{\sin \beta} = \frac{R_2}{\sin \theta} = \frac{R_1}{\sin \gamma} \tag{5.78}$$

The commutation current is

$$\overline{I_1} = R_2 \sin\gamma \tag{5.79}$$

Substituting (5.78) in (5.79) we find:

$$\overline{I_1} = \frac{R_1 \, R_2}{2} \sin \beta = \frac{R_1 \, R_2}{2} \sin\left(\pi - \frac{\pi}{\mu_o} \right) \tag{5.80}$$

The radius R_1 and R_2 are:

$$R_1 = \frac{\overline{I_1}}{2} \times \frac{\pi}{\mu_o} + 1 + q \tag{5.81}$$

$$R_2 = \frac{\overline{I_1}}{2} \times \frac{\pi}{\mu_o} + 1 - q \tag{5.82}$$

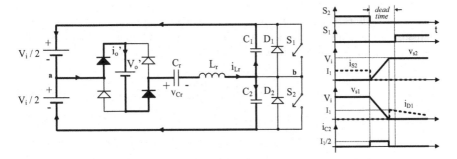

Fig. 5.16 Topological state and main waveforms for the second commutation time interval

Fig. 5.17 Commutation
current I_1 as a function of the
frequency ratio μ_o, taking the
static gain (q) as a parameter

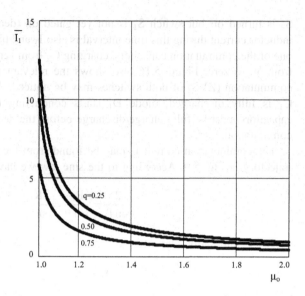

Substitution of (5.82) and (5.81) in (5.80) gives:

$$\overline{I_1} = \frac{1}{2} \times \left[\left(\frac{\overline{I_0}}{2} \times \frac{\pi}{\mu_o} + 1 \right)^2 - q^2 \right] \sin \left(\pi - \frac{\pi}{\mu_o} \right) \tag{5.83}$$

Substitution of (5.65) in (5.83) yields:

$$\overline{I_1} = \left(1 - q^2 \right) \times tg \left(\frac{\pi}{2 \, \mu_o} \right) \tag{5.84}$$

The normalized commutation current $\overline{I_1}$, as a function of the frequency ratio μ_o, taking the static gain (q) as a parameter, is plotted in Fig. 5.17. The smaller the commutation current, the longer the capacitors take to charge/discharge. It is not possible to ensure soft commutation in all load range because I_1 may reach a critical value, in which is not possible to fully charge/discharge the capacitors, and a dissipative commutation will take place. For a giving dead time and load range, the capacitors are calculated as follows:

$$C_1 = C_2 = \frac{I_1 t_d}{V_i} \tag{5.85}$$

5.5 Numerical Example

A series resonant converter that operates with the switching frequency above the resonant frequency, has the parameters given by Table 5.1.

Find: (a) the resonant inductance L_r, (b) the resonant capacitance C_r, (c) the capacitances of the commutation capacitors.

Solution:

The output DC voltage referred to the primary side of the transformer (V'_o) and the transformer turns ratio is calculated as follows:

$$V'_o = q\,V_1 = 0.6 \times \frac{400}{2} = 120\ V$$

$$n = \frac{N_1}{N_2} = \frac{V'_o}{V_o} = \frac{120}{50} = 2.4$$

At nominal power, the switching frequency is

$$f_{s\,min} = f_o\,\mu_o = 20 \times 10^3 \times 1.1 = 22 \times 10^3\ Hz$$

The resonant frequency is given by

$$f_o = \frac{1}{2\pi\sqrt{L_r\,C_r}} = 20 \times 10^3\ Hz$$

Therefore,

$$L_r\,C_r = 63.325 \times 10^{-12}$$

In the operation point, from the output characteristic we find the normalized output current:

Table 5.1 Specifications of the converter

Input DC voltage (V_i)	400 V
Output DC voltage (V_o)	50 V
Nominal average output current (I_o)	10 A
Nominal power (P_o)	500 W
Maximum switching frequency ($f_{s\,max}$)	40 kHz
Static gain (q)	0.6
Normalized switching frequency (μ_o)	1.1
Dead time (t_d)	1 μs

$$\overline{I}_o' = 3.25$$

The average output current referred to the primary side of the transformer gives a second ratio between the resonant components.

$$I_o' = \frac{I_o}{N_1/N_2} = \frac{10}{2.4} = 4.1667 \ A$$

From the equation

$$\overline{I}_o' = \frac{I_o' \sqrt{L_r/C_r}}{V_1}$$

we find

$$\frac{L_r}{C_r} = \left(\frac{\overline{I}_o' \ V_1}{I_o'}\right)^2 = 24,336$$

Thus, combining the previous results, we find:

$$C_r = 51 \ nF$$

$$L_r = 1.2414 \ mH$$

The series resonant converter operating above the switching frequency shown in Fig. 5.18 is simulate, with commutation capacitances $C_c = 2.125$ nF and a dead time of 455 ns. Figure 5.19 shows the voltage and current in the resonant tank, the AC voltage v_{ab} and the output current i_o'. The voltage and current in the switches are presented in Fig. 5.20 and a detail of the commutation showing the zero voltage switching (ZVS) in switch S_1 is presented in Fig. 5.21.

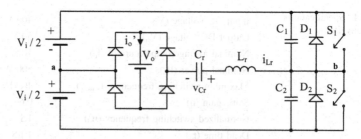

Fig. 5.18 Simulated circuit

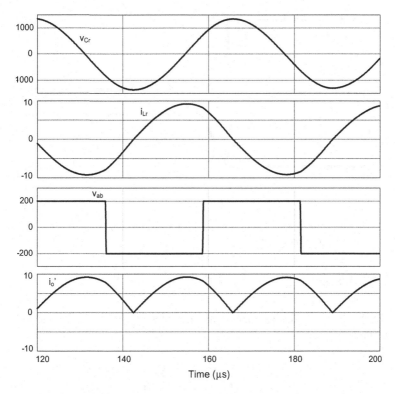

Fig. 5.19 Resonant capacitor voltage, resonant inductor current, AC voltage v_{ab} and output current i'_o, at nominal power

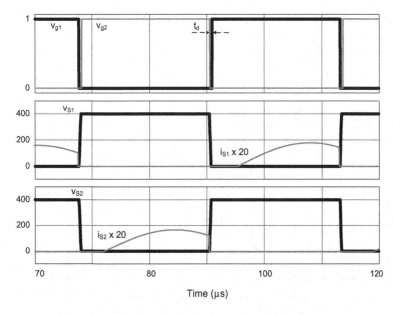

Fig. 5.20 Switches S_1 and S_2 commutation at nominal power: gate signals, voltage and current in the switches

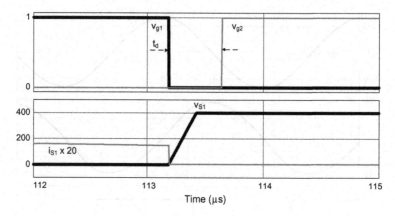

Fig. 5.21 A detail of switch S_1 commutation at turn off, for nominal power: S_1 and S_2 gate signals, voltage and current at the switch S_1

5.6 Problems

(1) The full-bridge series resonant converter shown in Fig. 5.22 has the following parameters:

$$Vi = 400 \text{ V} \qquad N_p/N_s = 1 \quad C_r = 52 \text{ nF}$$
$$L_r = 48.65 \text{ μH} \quad \mu_o = 1.5 \qquad R_o = 25 \text{ Ω}$$

Calculate:

(a) The resonant frequency;
(b) The output average current;
(c) The output voltage;
(d) The resonant capacitor peak voltage;
(e) The resonant inductor peak current.

   Answers: (a) $f_o = 100$ kHz;   (b) $I_o = 9.19$ A;   (c) $V_o = 229.7$ V;
(d) $V_{Co} = 294$ V; (e) 15.2 A.

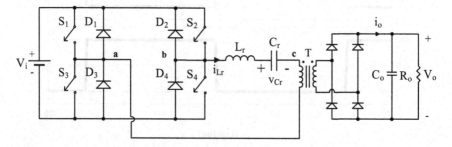

Fig. 5.22 Full-bridge series resonant converter

(2) Consider the full-bridge series resonant converter shown in Fig. 5.22 operating with $\mu_o = 1.5$ and the output short circuited. Find the equations for:

(a) The resonant capacitor peak voltage;
(b) The average output short circuit current;
(c) The resonant inductor peak current.

Answer: (a) $v_{Crpeak} = V_i$; (b) $I'_o = \frac{3}{\pi} \times \frac{V_i}{Z}$; (c) $i_{Lrpeak} = \frac{2}{Z} V_i$

(3) For the converter of previous exercises, find:

(a) The average output short circuit current;
(b) The resonant inductor peak current;
(c) The resonant capacitor peak voltage;
(d) The dc-bus average current.

Describe the converter operation, representing the topological states and the main waveforms, considering all components ideal and verify the results with simulation.

Answers: (a) $I_o = 12.4$ A; (b) $i_{Lrpeak} = 26.18$ A; (c) $v_{Crpeak} = 400$ V; (d) $i_{Vi} = 0$.

(4) The converter of exercise 1 operates with $\mu_o = 1.5$ and $V_o = 200$ V. Find the load resistance R_o.

Answer: $R_o = 20$ Ω.

Chapter 6
LLC Resonant Converter

6.1 Introduction

Figure 6.1 shows the power stage diagram of two common topological variations of the LLC converter, with a full bridge rectifier.

The Series Resonant Converter (SRC) is a particular topology of the LLC converter, where the magnetizing inductance is relatively large and not involved in the resonance operation [1–4].

The LLC converter has many benefits over the conventional series resonant converter. For example, it can regulate the output voltage over wide line and load variations with a relatively small variation of switching frequency, while preserving very high efficiency. It can also operate at zero voltage switching (ZVS) over the entire operating range.

A representation of the Full Bridge LLC converter with all parameters reflected to the primary side of the transformer, including the magnetizing inductance, is shown in Fig. 6.2.

For the converter's steady state analysis, the capacitor C_o in parallel with R_o is replaced with the voltage source V'_o.

In Fig. 6.2, it can be observed that there are two inductors (L_r and L_m) and a resonant capacitor (C_r) that provide unique characteristics, in comparison to others resonant converters, in terms of static and dynamic characteristics and also switching losses.

The LLC converter can operate with frequency modulation (FM) or pulse width modulation (PWM), but we will study only its behavior under frequency modulation, which is the most common case.

© Springer International Publishing AG, part of Springer Nature 2019
I. Barbi and F. Pöttker, *Soft Commutation Isolated DC-DC Converters*,
Power Systems, https://doi.org/10.1007/978-3-319-96178-1_6

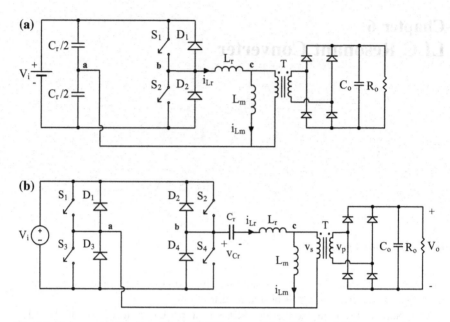

Fig. 6.1 Power stage diagram of the: **a** Half Bridge and **b** Full Bridge LLC resonant converter

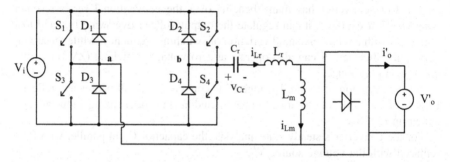

Fig. 6.2 Power stage of the Full Bridge LLC resonant converter, with ideal components and parameters reflected to the primary side of the transformer

6.2 First Harmonic Equivalent Circuit

Figure 6.3 represents the LLC converter in a different perspective with all parameters reflected to the transformer's primary side.

The waveform of the rectangular voltage V_x is represented in Fig. 6.4 with its fundamental harmonic. The converter switching frequency is represented by f_s.

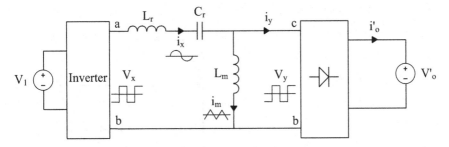

Fig. 6.3 Simplified representation of the LLC converter

Fig. 6.4 Waveform of the voltage V_x and its first harmonic

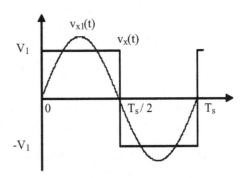

Then,

$$v_{x1}(t) = \frac{4V_1}{\pi}\sin(\omega_s t) \tag{6.1}$$

where

$$\omega_s = 2\pi f_s \tag{6.2}$$

and

$$T_s = \frac{1}{f_s} \tag{6.3}$$

The voltage V_y can also be considered rectangular, which amplitude is V_o'. Its waveform and first order harmonic are represented in Fig. 6.5.

The first harmonic of V_y is given by (6.4).

$$v_{y1}(t) = \frac{4V_o'}{\pi}\sin(\omega_s t - f) \tag{6.4}$$

The displacement angle between V_{x1} and V_{y1} is represented by ϕ.

Fig. 6.5 Waveform of the voltage V_y and its first harmonic

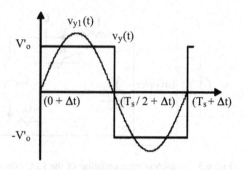

The first harmonic equivalent circuit of the LLC converter is shown in Fig. 6.6. The output diode rectifier keeps current i_{y1} and voltage V_{y1} in phase.

Applying Thevenin's Theorem, it can be obtained the equivalent circuit shown in Fig. 6.7.

The voltage V_{Tp} is defined by (6.5).

$$V_{Tp} = V_{xp} \frac{j\omega_s L_m}{\frac{1}{j\omega_s C_r} + j\omega_s(L_r + L_m)} \tag{6.5}$$

With the proper algebraic manipulation of (6.5) it can be obtained (6.6).

$$V_{Tp} = V_{xp} \frac{\omega_s^2 C_r L_m}{\omega_s^2 C_r(L_r + L_m) - 1} \tag{6.6}$$

The expression (6.6) shows that voltages V_{Tp} and V_{yp} are in phase.

Fig. 6.6 First harmonic equivalent circuit of the LLC converter

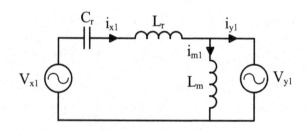

Fig. 6.7 Equivalent circuit obtained from Thevenin's Theorem

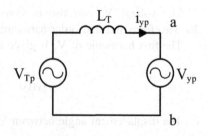

The equivalent Thevenin's impedance can be defined by (6.7).

$$Z_T = \frac{j\omega_s L_m \left(j\omega_s L_r + \frac{1}{j\omega_s C_r}\right)}{\frac{1}{j\omega_s C_r} + j\omega_s (L_r + L_m)} \tag{6.7}$$

With the proper algebraic manipulation of (6.7), (6.8) is obtained.

$$Z_T = \frac{j\omega_s L_m \left(\omega_s^2 C_r L_r - 1\right)}{\omega_s^2 C_r (L_r + L_m) - 1} \tag{6.8}$$

6.3 Voltage Gain Characteristics

To obtain the steady state curves of a LLC converter for sinusoidal currents and voltages, the phasorial representation is employed, as shown in Fig. 6.8.

$$V_{xp} = \frac{4V_1}{\pi} \tag{6.9}$$

$$V_{yp} = \frac{4V_o'}{\pi} \tag{6.10}$$

$$L_T = \frac{L_m \left(\omega_s^2 C_r L_r - 1\right)}{\omega_s^2 C_r (L_r + L_m) - 1} \tag{6.11}$$

In Fig. 6.8, the phasors represent voltages and currents amplitude. The angle ϕ represents the phase displacement between the voltages V_{xp} and V_{yp}. From the analysis of the phasorial diagram shown in Fig. 6.8, (6.12) can be written.

$$V_{Tp}^2 = V_{yp}^2 + \left(\omega_s L_T I_{yp}\right)^2 \tag{6.12}$$

Thus:

$$V_{Tp}^2 = \left(\frac{4V_o'}{\pi}\right)^2 + \left(\omega_s L_T I_{yp}\right)^2 \tag{6.13}$$

Fig. 6.8 Phasorial representation for the first harmonic circuit

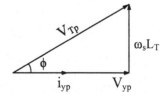

The peak value of I_y, in terms of the average value of the reflected load to the primary winding I_o, is given by (6.14).

$$I_{yp} = \frac{\pi}{2}I_o'$$

(6.14)

Substituting (6.14) into (6.13), the expression (6.15) can be obtained.

$$V_{Tp}^2 = \left(\frac{4V_o'}{\pi}\right)^2 + \left(\frac{\omega_s L_T \pi I_o'}{2}\right)^2$$

(6.15)

From the substitution of (6.6) and (6.11) into (6.15), (6.16) can be written.

$$\left(\frac{4V_1}{\pi}\right)^2\left[\frac{\omega_s^2 C_r L_m}{\omega_s^2 C_r(L_r + L_m) - 1}\right]^2 = \left(\frac{4V_o'}{\pi}\right)^2 + \left(\frac{\pi}{2}\right)^2\left[\frac{\omega_s L_m\left(\omega_s^2 C_r L_r - 1\right)I_o'}{\omega_s^2 C_r(L_r + L_m) - 1}\right]^2$$

(6.16)

The simplification of (6.16) leads to (6.17).

$$(V_o')^2 = (V_1)^2\left[\frac{\omega_s^2 C_r L_m}{\omega_s^2 C_r(L_r + L_m) - 1}\right]^2 - \frac{\pi^4}{64}\left[\frac{\omega_s L_m\left(\omega_s^2 C_r L_r - 1\right)I_o'}{\omega_s^2 C_r(L_r + L_m) - 1}\right]^2$$

(6.17)

Considering the following definitions:

$$M = \frac{V_o'}{V_1}$$

(6.18)

$$\omega_s = 2\pi f_s$$

(6.19)

$$\omega_{r1} = \frac{1}{\sqrt{C_r L_r}}$$

(6.20)

$$\omega_n = \frac{\omega_s}{\omega_{r1}}$$

(6.21)

Therefore:

$$\omega_s^2 C_r L_r = \omega_n^2$$

(6.22)

$$\lambda = \frac{L_r}{L_m}$$

(6.23)

$$\bar{I}_o' = \frac{\omega_s L_r}{V_1}I_o'$$

(6.24)

Replacing (6.18) and (6.24) into (6.17), it can obtain the expression (6.25).

$$M^2 = \left[\frac{\omega_n^2}{\omega_n^2(\lambda + 1) - \lambda} \right]^2 - \frac{\pi^4}{64} \left[\frac{(\omega_n^2 - 1)\overline{I_o'}}{\omega_n^2(\lambda + 1) - \lambda} \right]^2 \tag{6.25}$$

Appropriate algebraic manipulation gives

$$M = \frac{\sqrt{\omega_n^4 - \frac{\pi^4}{64}\left[(\omega_n^2 - 1)\overline{I_o'}\right]^2}}{\omega_n^2(\lambda + 1) - \lambda} \tag{6.26}$$

Also:

$$\omega_{r1} = 2\pi f_{r1} \tag{6.27}$$

$$\omega_s = 2\pi f_s \tag{6.28}$$

In addition:

$$\omega_n = \frac{f_s}{f_{r1}} \tag{6.29}$$

$$\omega_n = f_n \tag{6.30}$$

Then, the voltage gain M can now be represented by (6.31).

$$M = \frac{\sqrt{f_n^4 - \frac{\pi^4}{64}\left[(f_n^2 - 1)\overline{I_o'}\right]^2}}{f_n^2(\lambda + 1) - \lambda} \tag{6.31}$$

When $f_r = f_s$, the relative frequency $f_n = 1$ and $M = 1$, regardless the value of the current $\overline{I_o'}$. The resonance frequency f_r is defined by the values of C_r and L_r, according to the expression (6.32).

$$f_{r1} = \frac{1}{2\pi\sqrt{C_r L_r}} \tag{6.32}$$

The converter's voltage gain is infinite, theoretically, when the denominator of (6.32), represented by (6.33), is equals to zero.

$$f_n^2(\lambda + 1) - \lambda = 0 \tag{6.33}$$

Therefore, when:

$$f_n = \sqrt{\frac{\lambda}{\lambda+1}} \qquad (6.34)$$

Substituting (6.23) into (6.34), leads to (6.35).

$$f_n = \sqrt{\frac{L_r}{L_r + L_m}} \qquad (6.35)$$

Also:

$$f_n = \frac{f_s}{f_{r1}} \qquad (6.36)$$

Therefore:

$$f_s = f_r \sqrt{\frac{L_r}{L_r + L_m}} \qquad (6.37)$$

Substituting (6.32) into (6.37) leads to (6.38).

$$f_s = \frac{1}{2\pi\sqrt{C_r L_r}} \sqrt{\frac{L_r}{L_r + L_m}} \qquad (6.38)$$

Thus:

$$f_s = \frac{1}{2\pi\sqrt{C_r(L_r + L_m)}} \qquad (6.39)$$

The Eq. (6.39) represents the resonance frequency of a series resonant circuit, composed by C_r, L_r and L_m, as shown in (6.40).

$$f_{r2} = \frac{1}{2\pi\sqrt{C_r(L_r + L_m)}} \qquad (6.40)$$

Dividing (6.40) by (6.32), the relation (6.41) is obtained.

$$\frac{f_{r2}}{f_{r1}} = \sqrt{\frac{\lambda}{\lambda+1}} \qquad (6.41)$$

Equation (6.41) is also graphically represented for Fig. 6.9.

From the design step of the LLC resonant converter, it can be ensured that f_{r2} is sufficiently smaller than f_{r1}. A typical value for λ is 0.2, which implies in $f_{r2} = 0.4f_{r1}$.

Fig. 6.9 f_s/f_r for different
values of λ

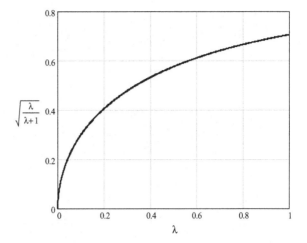

Fig. 6.10 Voltage gain M for
different values of the
parametrized current $\overline{I}'_o$

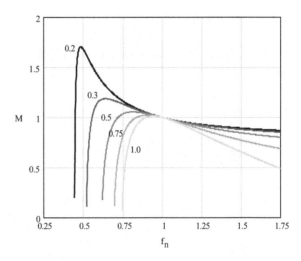

The curves relating M and the normalized frequency f_n for different values of the parametrized current $\overline{I}'_o$ are shown in Fig. 6.10. The ratio λ between the inductances is 0.2.

Figure 6.11 represents the static gain M as function of the parametrized current $\overline{I}'_o$ for different values of relative frequency f_n.

It can be noticed from these curves that the converter's gain is controlled by switching frequency. The load voltage reflected to the primary winding of the transformer can be higher or lower than the input voltage, for switching frequency greater or smaller than the resonance frequency f_{r1}, respectively. The curves also show that for $0 \leq \overline{I}'_o \leq 0.5$ and $0.8 \leq f_n \leq 1.2$, the static gain is not very sensitive to the load current.

Fig. 6.11 Voltage gain M versus $\overline{I'_o}$ for different values of f_n

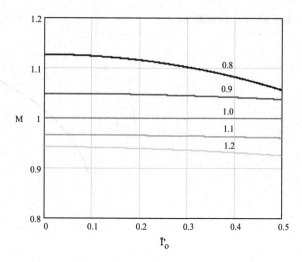

6.3.1 Numerical Example 1

A LLC converter, with the transformer turns ratio equals to one, has the following parameters:

- V_1 = 400 V
- L_r = 14 μH
- L_m = 70 μH
- C_r = 47 nF
- f_s = 170 kHz
- I_o = 5 A

Find the voltage and the power in the load.
Solution

$$\lambda = \frac{L_r}{L_m} = \frac{14 \times 10^{-6}}{70 \times 10^{-6}} = 0.2$$

$$f_{r1} = \frac{1}{2\pi\sqrt{L_r C_r}} = \frac{1}{2\pi\sqrt{14 \times 10^{-6} \times 47 \times 10^{-9}}} = 196.2\,\text{kHz}$$

$$\omega_{r1} = 2\pi f_{r1} = 1{,}232{,}748\,\text{rad/s}$$

$$f_n = \frac{f_s}{f_{r1}} = \frac{170{,}000}{196{,}204} = 0.866$$

$$\overline{I'_o} = \frac{\omega_{r1} L_r}{V_1} I_o = \frac{1{,}232{,}748 \times 14 \times 10^{-6}}{400} \times 5 = 0.216$$

$$M = \frac{\sqrt{f_n^4 - \frac{\pi^4}{64}\left[(f_n^2 - 1)\overline{I_o'}\right]^2}}{f_n^2(\lambda + 1) - \lambda} = 1.067$$

$$\boxed{V_o' = MV_1 = 1.067 \times 400 = 426.78\,\text{V}}$$

$$\boxed{P_o = V_o'I_o' = 426.78 \times 5 = 2134\,\text{W}}$$

6.4   Static Gain Without Load (Open Circuit)

When the converter operates without load, the processed power is zero. In this case, the parametrized load current is also zero. Consequently, the static gain, given by (6.42) and shown in Fig. 6.12, can be deduced from the gain expression given by (6.31).

$$M = \frac{f_n^2}{f_n^2(\lambda + 1) - \lambda} \tag{6.42}$$

It is observed that the static gain is greater than one when $f_n < 1$ and smaller than one when $f_n > 1$. The converter operates properly without load and can have its voltage controlled through the switching frequency.

This is one of the most important properties of the LLC converter, and one of its advantages over the series and parallel resonant converters, which do not have these properties.

Fig. 6.12 Static gain of a LLC without load for different values of λ

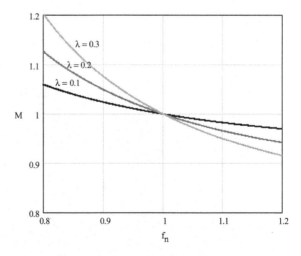

6.5 Static Gain as a Function of Load Resistance

The expression (6.31), reproduced from now on as (6.43), represents the static gain as a function of the parameterized value of the average load current reflected to the primary side of the transformer.

$$M = \frac{\sqrt{f_n^4 - \left[\frac{\pi^2}{8}\left(f_n^2 - 1\right)\overline{I}_o'\right]^2}}{f_n^2(\lambda + 1) - \lambda} \tag{6.43}$$

The parameterized current $\overline{I}_o'$ is defined by

$$\overline{I}_o' = \frac{\omega_s L_r}{V_1} I_o' \tag{6.44}$$

However:

$$I_o' = \frac{V_o'}{R_o'} \tag{6.45}$$

The parameter R_o' is defined as the load resistance reflected to the primary side of the transformer.

Therefore:

$$\overline{I}_o' = \frac{\omega_s L_r}{V_1} \frac{V_o'}{R_o'} \tag{6.46}$$

Since

$$M = \frac{V_o'}{V_1} \tag{6.47}$$

We obtain:

$$\overline{I}_o' \frac{\pi^2}{8} = M \frac{\omega_n \frac{1}{\sqrt{C_r L_r}} L_r}{R_o' \frac{8}{\pi^2}} \tag{6.48}$$

Defining

$$R_{ac} = R_o' \frac{8}{\pi^2} \tag{6.49}$$

And

$$Q = \frac{\sqrt{\frac{L_r}{C_r}}}{R_{ac}} \tag{6.50}$$

R_{ac} represents the load resistance reflected to the primary winding of the transformer for sinusoidal current, and Q the quality factor.
Thus:

$$\overline{I'_o} \frac{\pi^2}{8} = f_n MQ \tag{6.51}$$

Substitution of (6.51) into (6.43), with appropriate algebraic manipulation, gives

$$M = \frac{f_n^2}{\sqrt{\left[f_n^2(\lambda + 1) - \lambda\right]^2 + \left[f_n Q(f_n^2 - 1)\right]^2}} \tag{6.52}$$

Figure 6.13 is shown the static gain (M) given by Eq. (6.52), as function of f_n, for different values of the quality factor (Q).

Figure 6.14 shows the static gain (M) as a function of Q for different values of f_n.

When the output power is zero, $Q = 0$ and the static gain expression is represented by (6.53). This same equation was found in (6.42) considering the parametrized load current as zero, which also implies in no power transferred to the load.

$$M = \frac{f_n^2}{f_n^2(\lambda + 1) - \lambda} \tag{6.53}$$

Fig. 6.13 Static gain as a function of f_n for different values of Q

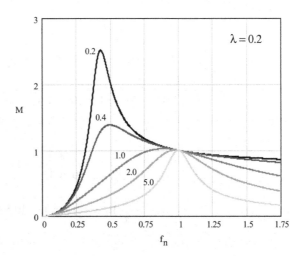

Fig. 6.14 Static gain as function of Q for different values of f_n

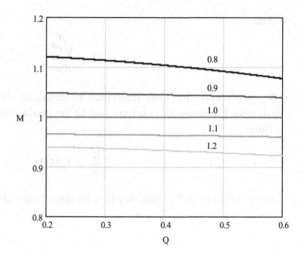

6.5.1 Numerical Example 2

A LLC converter, with transformer turns ratio a = 1, operates with the following parameters:

- V_1 = 400 V
- L_r = 14 µH
- L_m = 70 µH
- C_r = 47 nF
- f_s = 220 kHz
- R_o = 50 Ω

Find the values of the voltage and the power in the load resistor.
Solution

$$\lambda = \frac{L_r}{L_m} = \frac{14 \times 10^{-6}}{70 \times 10^{-6}} = 0.2$$

$$f_{rl} = \frac{1}{2\pi\sqrt{L_rC_r}} = \frac{1}{2\pi\sqrt{14 \times 10^{-6} \times 47 \times 10^{-9}}} = 196.2\,\text{kHz}$$

$$\omega_{rl} = 2\pi f_{rl} = 1,232,748\,\text{rad/s}$$

$$f_n = \frac{f_s}{f_{rl}} = \frac{220,000}{196,204} = 1.121$$

$$R_{ac} = \frac{8}{\pi^2}R_o = \frac{8}{\pi^2} \times 50 = 40.53\,\Omega$$

$$Q = \frac{\sqrt{\frac{L_r}{C_r}}}{R_{ac}} = \frac{\sqrt{\frac{14\times10^{-6}}{47\times10^{-9}}}}{40.53} = 0.426$$

$$M = \frac{f_n^2}{\sqrt{\left[f_n^2(\lambda+1) - \lambda\right]^2 + \left[f_n Q(f_n^2 - 1)\right]^2}} = 0.956$$

$$\boxed{V_o' = MV_1 = 0.956 \times 400 = 382.59 \text{ V}}$$

$$\boxed{P_o = \frac{(V_o')^2}{R_o} = \frac{382.59^2}{50} = 2928 \text{ W}}$$

6.6   Minimum Normalized Frequency

According to the curves plotted in Fig. 6.15, for each value of the parametrized current $\overline{I_o'}$, the static gain (M) as a function of normalized frequency f_n has a peak value that defines the boundary between capacitive or inductive input impedance of the LLC converter. The capacitive impedance happens when the converter operates below the minimum allowed switching frequency and it must be avoided.

For a given value of $\overline{I_o'}$, at $f_n < f_{nmin}$, the input impedance becomes capacitive and when $f_n > f_{nmin}$, it becomes inductive.

To ensure ZVS operation, it is necessary an inductive impedance, so the current is delayed in relation to the voltage V_{ab}. The minimum normalized frequency can be found by differentiation of (6.31) in relation to f_n, as follows.

Fig. 6.15 Static gain of a LLC converter for different values of $\overline{I_o'}$

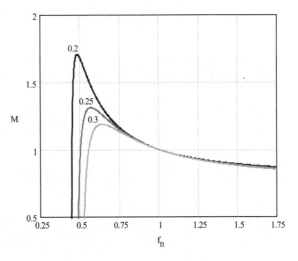

$$\frac{dM\left(\overline{I}_o', \lambda, f_n\right)}{df_n} = \frac{-2f_n\left(\alpha\overline{I}_o'^2 f_n^2 - \alpha\overline{I}_o'^2 + \lambda f_n^2\right)}{\left(\lambda f_n^2 - \lambda + f_n^2\right)^2 \sqrt{f_n^4 - \alpha\overline{I}_o'^2\left(f_n^2 - 1\right)^2}} \tag{6.54}$$

where:

$$\alpha = \frac{\pi^4}{64} \tag{6.55}$$

To make (6.54) equals to zero,

$$\frac{dM\left(\overline{I}_o', \lambda, f_n\right)}{df_n} = 0 \tag{6.56}$$

it is necessary that

$$\alpha\overline{I}_o'^2 f_n^2 - \alpha\overline{I}_o'^2 + \lambda f_n^2 = 0 \tag{6.57}$$

Therefore:

$$f_{nmin} = \sqrt{\frac{\alpha\overline{I}_o'^2}{\alpha\overline{I}_o'^2 + \lambda}} \tag{6.58}$$

Then:

$$f_{nmin} = \frac{1}{\sqrt{1 + \frac{64\lambda}{\pi^4 \overline{I}_o'^2}}} \tag{6.59}$$

Expression (6.59) is plotted in Fig. 6.16. The value of f_{min} for a given value of λ is determined for operations with maximum power or for the maximum value of parametrized load current $\overline{I}_o'$.

For a given value of $\overline{I}_o'$, the value of f_{min} will be naturally respected.

Fig. 6.16 Minimum
normalized frequency as a
function of parametrized
load current for different
values of λ

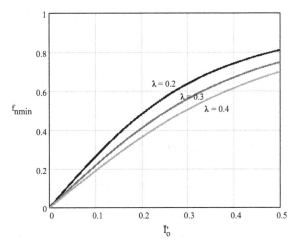

6.6.1   *Numerical Example 3*

A LLC converter operates with the following parameters:

- $V_1 = 400$ V
- $L_r = 14$ μH
- $L_m = 70$ μH
- $C_r = 47$ nF
- $a = 1$
- $I_o = 5$ A (maximum value)

Find the minimum operating frequency required to the LLC input reactance be
inductive.

Solution

$$\lambda = \frac{L_r}{L_m} = \frac{14 \times 10^{-6}}{70 \times 10^{-6}} = 0.2$$

$$f_{r1} = \frac{1}{2\pi\sqrt{L_r C_r}} = \frac{1}{2\pi\sqrt{14 \times 10^{-6} \times 47 \times 10^{-9}}} = 196.2 \, \text{kHz}$$

$$\omega_{r1} = 2\pi f_{r1} = 1{,}232{,}748 \, \text{rad/s}$$

$$\overline{I'_{omax}} = \frac{\omega_{r1} L_r}{V_1} I_{omax} = \frac{1{,}232{,}748 \times 14 \times 10^{-6}}{400} \times 5 = 0.216$$

$$f_{nmin} = \frac{1}{\sqrt{1 + \frac{64\lambda}{\pi^4 I'^2_{omax}}}} = 0.511$$

$$\boxed{f_{smin} = f_{nmin}f_{r1} = 0.511 \times 196.2 = 100.34\,\text{kHz}}$$

6.7 Commutation's Analysis

In order to obtain ZVS commutation for the entire range of input voltage and output power, it is necessary that the switching capacitors be fully charged or discharged during the "dead time".

Therefore, it is necessary that the current in the switch at the instant it is turned off be higher than a minimum value needed to completely charge the commutation equivalent capacitor.

The lowest current value occurs at the maximum operating frequency with no load.

Figure 6.17 shows the equivalent circuit of the LLC with no load.

The typical waveforms of the LLC operating at no load are shown in Fig. 6.18. At the instant that the switches turn off, the value of the current responsible for charging and discharging the switching capacitor is I_{mp}, which will be determined as follows.

Fig. 6.17 Equivalent circuit of a LLC converter without load

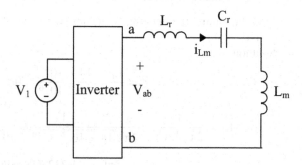

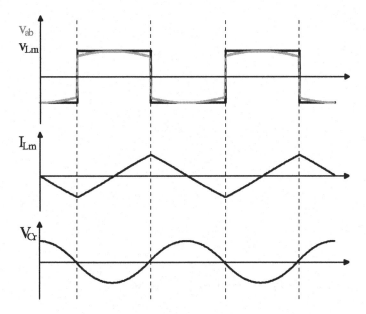

Fig. 6.18 Typical waveforms of a LLC operating without load

6.7.1 Current in the Switches During Commutation

During the interval [0, $T_s/2$] the current is represented by Eq. (6.60).

$$i_{Lm}(t) = \frac{V_1}{Z_2}\sin(\omega_2 t) - I_{mp}\cos(\omega_2 t) \qquad (6.60)$$

where:

$$Z_2 = \sqrt{\frac{L_r + L_m}{C_r}} \qquad (6.61)$$

and

$$\omega_2 = \frac{1}{\sqrt{C_r(L_r + L_m)}} \qquad (6.62)$$

Then:

$$\frac{Z_2 i_{Lm}(t)}{V_1} = \sin(\omega_2 t) - \frac{Z_2 I_{mp}}{V_1}\cos(\omega_2 t) \qquad (6.63)$$

Defining:

$$\overline{i_{Lm}}(t) = \frac{Z_2 i_{Lm}(t)}{V_1} \tag{6.64}$$

and

$$\overline{I_{mp}} = \frac{Z_2 I_{mp}}{V_1} \tag{6.65}$$

Then:

$$\overline{i_{Lm}}(t) = \sin(\omega_2 t) - \overline{I_{mp}} \cos(\omega_2 t) \tag{6.66}$$

At the instant t = T/2, $\overline{i_{Lm}}(t) = \overline{I_{mp}}$.
Then:

$$\overline{i_{Lm}}(t) = \sin\left(\omega_2 \frac{T}{2}\right) - \overline{I_{mp}} \cos\left(\omega_2 \frac{T}{2}\right) \tag{6.67}$$

However:

$$\frac{T}{2} = \frac{1}{2f_s} \tag{6.68}$$

Thus:

$$\omega_2 \frac{T}{2} = 2\pi f_{r2} \frac{1}{2f_s} \tag{6.69}$$

$$\omega_2 \frac{T}{2} = \frac{\pi f_{r2}}{f_s} \tag{6.70}$$

Substituting (6.70) into (6.66) and manipulating algebraically, we obtain:

$$\overline{I_{mp}} = \frac{\sin\left(\frac{\pi f_{r2}}{f_s}\right)}{1 + \cos\left(\frac{\pi f_{r2}}{f_s}\right)} \tag{6.71}$$

It was previously determined that:

$$f_{r2} = f_{r1} \sqrt{\frac{\lambda}{\lambda + 1}} \tag{6.72}$$

Thus:

$$\frac{f_{r2}}{f_s} = \frac{f_{r1}}{f_s} \sqrt{\frac{\lambda}{\lambda+1}} \tag{6.73}$$

With:

$$\frac{f_s}{f_{r1}} = f_n \tag{6.74}$$

It is obtained:

$$\frac{f_{r2}}{f_s} = \frac{1}{f_n} \sqrt{\frac{\lambda}{\lambda+1}} \tag{6.75}$$

Substituting (6.75) into (6.71) gives (6.76).

$$\overline{I_{mp}} = \frac{\sin\left(\frac{\pi}{f_n}\sqrt{\frac{\lambda}{\lambda+1}}\right)}{1 + \cos\left(\frac{\pi}{f_n}\sqrt{\frac{\lambda}{\lambda+1}}\right)} \tag{6.76}$$

Equation (6.76) represents the parametrized current in the switch at the instant it is turned-off, as a function of the normalized switching frequency f_n and the ratio of the resonant magnetizing inductance λ.

In Fig. 6.19 the parameterized current $\overline{I_{mp}}$ is plotted as a function of the normalized switching frequency f_n, for $\lambda = 0.2$.

Fig. 6.19 Parameterized current $\overline{I_{mp}}$ as a function of the normalized switching frequency

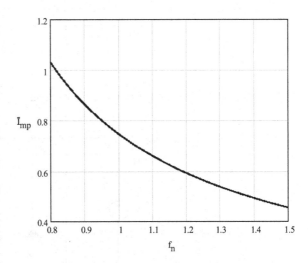

With appropriate algebraic manipulation of Eq. (6.76), Eq. (6.77) can be obtained.

$$\overline{I_{mp}} = \tan\left(\frac{\pi}{2f_n}\sqrt{\frac{\lambda}{\lambda+1}}\right) \tag{6.77}$$

Figure 6.19 shows that the current at the switching instant decreases with the increasing of the switching frequency.

6.7.2   Commutation Analysis

Figure 6.20 shows the topological states of a commutation leg of the converter during a commutation time interval.

The corresponding relevant waveforms are shown in Fig. 6.21.

During the topological state "a", before the commutation stage begins, S_1 conducts current I_{mp}, $I_{S2} = 0$, $V_{C1} = 0$ and $V_{C2} = V_1$.

Fig. 6.20 Topological states of a converter's leg during a commutation interval

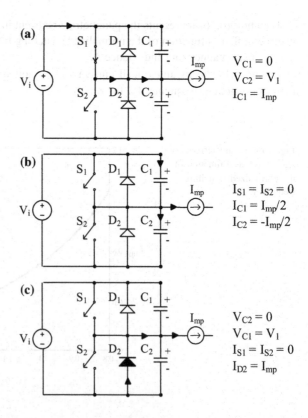

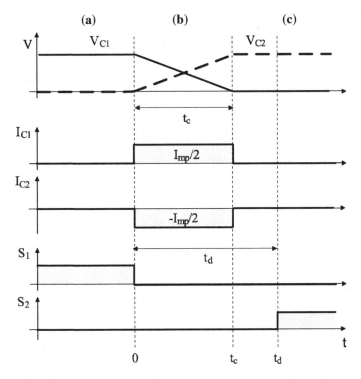

Fig. 6.21 Relevant waveforms during a commutation interval

During the topological state "b", after turning off the switch S_1 at the instant $t = 0$ s, the current I_{mp} charges the capacitor C_1, whose voltage rises from zero to V_i, while the capacitor C_2 is completely discharged. Since I_{mp} remains constant during the switching interval, V_{C1} rises and V_{C2} falls linearly. This results in zero voltage turn-off, since the voltage across the switch rises while its current is zero.

At the instant $t = t_c$, the current I_{mp} starts circulating through diode D_2. At the instant $t = t_d$, the switch S_2 is gated on at zero voltage and current.

Since:

$$V_{C1}(t) + V_{C2}(t) = V_1 \tag{6.78}$$

$$\frac{dV_{C1}(t)}{dt} + \frac{dV_{C2}(t)}{dt} = 0 \tag{6.79}$$

Assuming that $C_1 = C_2 = C$. Then:

$$C\frac{dV_{C1}(t)}{dt} + C\frac{dV_{C2}(t)}{dt} = 0 \tag{6.80}$$

However:

$$i_{C1}(t) = C \frac{dV_{C1}(t)}{dt} \tag{6.81}$$

and

$$i_{C2}(t) = C \frac{dV_{C1}(t)}{dt} \tag{6.82}$$

As a consequence:

$$i_{C2}(t) = -i_{C1}(t) \tag{6.83}$$

Since:

$$i_{C1}(t) + i_{C2}(t) = I_{mp} \tag{6.84}$$

We obtain:

$$i_{C1}(t) = \frac{I_{mp}}{2} \tag{6.85}$$

$$i_{C2}(t) = -\frac{I_{mp}}{2} \tag{6.86}$$

Since:

$$i_{C2}(t) = C \frac{V_1}{t} \tag{6.87}$$

Defining:

$$\frac{I_{mp}}{2} = C \frac{V_1}{t_c} \tag{6.88}$$

Therefore, for the complete discharge of C_2, it is necessary that:

$$I_{mp} \geq 2C \frac{V_1}{t_d} \tag{6.89}$$

According to (6.77):

$$\overline{I_{mp}} = \tan\left(\frac{\pi}{2f_n} \sqrt{\frac{\lambda}{\lambda+1}}\right) \tag{6.90}$$

With:

$$\overline{I_{mp}} = \frac{Z_2 I_{mp}}{V_i} \tag{6.91}$$

$$I_{mp2} = \frac{V_1}{Z_2} \overline{I_{mp}} \tag{6.92}$$

Or also:

$$I_{mp} = \frac{V_1}{Z_2} \tan\left(\frac{\pi}{2f_n} \sqrt{\frac{\lambda}{\lambda+1}}\right) \tag{6.93}$$

Combining (6.93) and (6.89) leads to (6.94):

$$\frac{V_1}{Z_2} \tan\left(\frac{\pi}{2f_n} \sqrt{\frac{\lambda}{\lambda+1}}\right) \geq 2C \frac{V_1}{t_d} \tag{6.94}$$

Or also:

$$\tan\left(\frac{\pi}{2f_n} \sqrt{\frac{\lambda}{\lambda+1}}\right) \geq 2C \frac{Z_2}{t_d} \tag{6.95}$$

where:

$$Z_2 = \sqrt{\frac{L_r + L_m}{C_r}} \tag{6.96}$$

6.7.3 Numerical Example 4

Find the needed gate signal dead-time for the LLC converter shown in Fig. 6.22 operating with no load and with the following parameters:

- $V_1 = 400$ V
- $L_r = 14$ μH
- $L_m = 70$ μH
- $C_r = 47$ nF
- $C = 2.2$ nF ($C_1 = C_2 = C_3 = C_4 = C$)
- $f_n = 1$

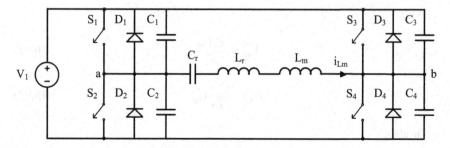

Fig. 6.22 Ideal equivalent LLC converter without load

Solution

$$\lambda = \frac{L_r}{L_m} = \frac{14 \times 10^{-6}}{70 \times 10^{-6}} = 0.2$$

$$f_{rl} = \frac{1}{2\pi\sqrt{L_rC_r}} = \frac{1}{2\pi\sqrt{14 \times 10^{-6} \times 47 \times 10^{-9}}} = 196.2\,\text{kHz}$$

$$\overline{I_{mp}} = \tan\left(\frac{\pi}{2f_n}\sqrt{\frac{\lambda}{\lambda+1}}\right) = 0.747$$

$$Z_2 = \sqrt{\frac{L_r + L_m}{C_r}} = 42.27\,\Omega$$

$$I_{mp} = \frac{V_1}{Z_2}\overline{I_{mp}} = \frac{400}{42.27} \times 0.747 = 7.06\,\text{A}$$

$$\boxed{t_c = 2C\frac{V_1}{I_{mp}} \approx 250\,\text{ns}}$$

The ZVS is achieved, provided that the dead time between the gate signals of the switches of a leg, t_d, is larger than t_c. Therefore:

$$t_d > 250\,\text{ns}$$

Figure 6.23 shows the relevant waveforms obtained by simulation, during the turning off of switch S_1. All components were considered ideal and a $t_d = 400$ ns was adopted. During the commutation time interval, the I_{Lr} current remains practically constant, equals to I_{mp}. Due to the high frequency operation, the dead time t_d represents a significant portion of the switching period, and therefore has some impact on the converter static gain.

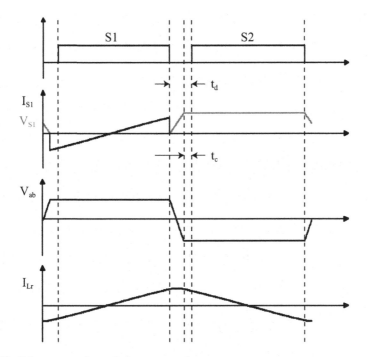

Fig. 6.23 Relevant waveforms during commutation

6.8 Input Impedance

Figure 6.24 shows the equivalent circuit of the LLC converter for the fundamental harmonics of voltage and current.

The input impedance is given by (6.97).

$$Z_1 = \frac{j\omega_s L_m R}{R + j\omega_s L_m} + j\omega_s L_r + \frac{1}{j\omega_s C_r} \tag{6.97}$$

We know that:

$$\lambda = \frac{L_r}{L_m} \tag{6.98}$$

Fig. 6.24 Equivalent LLC converter circuit for first order harmonic

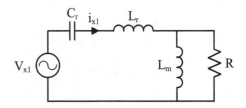

$$f_{rl} = \frac{1}{2\pi\sqrt{C_r L_r}} \tag{6.99}$$

$$Q = \frac{\omega_r L_r}{R} \tag{6.100}$$

$$f_n = \frac{f_s}{f_{rl}} \tag{6.101}$$

Substituting (6.98) to (6.101) into (6.96), leads to (6.102),

$$\overline{Z_1}(\lambda, f_n, Q) = \overline{R_1}(\lambda, f_n, Q) + j\overline{X_1}(\lambda, f_n, Q) \tag{6.102}$$

where

$$\overline{R_1}(\lambda, f_n, Q) = \frac{Q f_n^2}{Q^2 f_n^2 + \lambda^2} \tag{6.103}$$

and

$$\overline{X_1}(\lambda, f_n, Q) = \frac{f_n^2 - 1}{f_n} + \frac{\lambda f_n}{Q^2 f_n^2 + \lambda^2} \tag{6.104}$$

$\overline{Z_1}(\lambda, f_n, Q)$, $\overline{R_1}(\lambda, f_n, Q)$, $\overline{R_1}(\lambda, f_n, Q)$ and $\overline{X_1}(\lambda, f_n, Q)$ are defined as follows:

$$\overline{Z_1}(\lambda, f_n, Q) = \frac{Z_1(\lambda, f_n, Q)}{Z_{rl}} \tag{6.105}$$

$$\overline{R_1}(\lambda, f_n, Q) = \frac{R_1(\lambda, f_n, Q)}{R_{rl}} \tag{6.106}$$

$$\overline{X_1}(\lambda, f_n, Q) = \frac{X_1(\lambda, f_n, Q)}{X_{rl}} \tag{6.107}$$

$$Z_{rl} = \sqrt{\frac{L_r}{C_r}} \tag{6.108}$$

Therefore, the impedance Z_1 and its components R_1 and X_1 have been parameterized to the characteristic impedance Z_{rl}.

Figure 6.25 shows curves of the module of the parametrized $\overline{Z_1}$ as a function of f_n for $\lambda = 0.2$ and different values of Q.

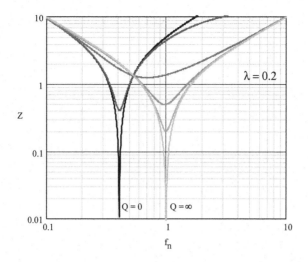

Fig. 6.25 Normalized input impedance as a function of f_n for different values of Q

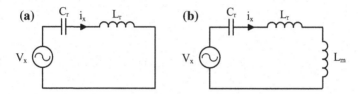

Fig. 6.26 Equivalent circuits of the LLC converter at: **a** output short-circuit and **b** output open-load

The output short-circuit condition implies $R = 0$ and $Q = \infty$. The corresponding equivalent circuit is shown in Fig. 6.26a. The resonance occurs for $f_n = 1$.

The output open-circuit, on the other hand, implies in $R = \infty$ and $Q = 0$. The corresponding equivalent circuit is shown in Fig. 6.26b and the normalized resonant frequency in given by (6.109).

$$f_{n2} = \sqrt{\frac{\lambda}{1 + \lambda}} \tag{6.109}$$

All impedance curves intersect themselves at the normalized frequency given by (6.110).

$$f_{n3} = \sqrt{\frac{2\lambda}{1 + 2\lambda}} \tag{6.110}$$

6.9   Resonant Capacitor Current and Voltage

The peak current through C_r is given by (6.111).

$$I_{xp} = \frac{V_{xp}}{|Z_1|} \tag{6.111}$$

However:

$$V_{xp} = \frac{4V_1}{\pi} \tag{6.112}$$

and

$$|Z_1| = Z_{r1}|\overline{Z_1}| \tag{6.113}$$

Then:

$$I_{xp} = \frac{4V_1}{\pi} \frac{1}{Z_{r1}|\overline{Z_1}|} \tag{6.114}$$

Thus:

$$\frac{Z_{r1}I_{xp}}{V_1} = \frac{4}{\pi|\overline{Z_1}|} \tag{6.115}$$

The parameterized current can be defined as (6.116).

$$\overline{I_{xp}} = \frac{Z_{r1}I_{xp}}{V_1} \tag{6.116}$$

Therefore:

$$\overline{I_{xp}} = \frac{4}{\pi|\overline{Z_1}|} \tag{6.117}$$

Or also:

$$\overline{I_{xp}}(\lambda, f_n, Q) = \frac{4}{\pi|\overline{Z_1}(\lambda, f_n, Q)|} \tag{6.118}$$

Fig. 6.27 Parametrized peak current through C_r as a function of f_n for different values of Q

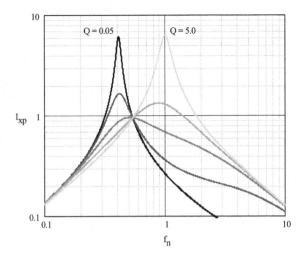

where:

$$|\overline{Z_1}(\lambda, f_n, Q)| = \sqrt{\left(\frac{Qf_n^2}{Q^2f_n^2 + \lambda^2}\right)^2 + \left(\frac{f_n^2 - 1}{f_n} + \frac{\lambda f_n}{Q^2f_n^2 + \lambda^2}\right)^2} \qquad (6.119)$$

Figure 6.27 shows the parametrized peak current through the capacitor C_r, as a function of the normalized switching frequency f_n, for different values of the quality factor Q.

Like what occurs with the input impedance, all curves of the current intersect at $f_n = f_{n3}$.

The peak voltage across the resonant capacitor is defined by (6.120).

$$V_{crp} = \frac{I_{xp}}{\omega_s C_r} \qquad (6.120)$$

Defining:

$$\overline{V_{crp}} = \frac{V_{crp}}{V_1} \qquad (6.121)$$

Then:

$$\overline{V_{crp}} = \frac{I_{xp}}{V_1 \omega_s C_r} \qquad (6.122)$$

Fig. 6.28 Parametrized peak
voltage across the resonant
capacitor C_r as a function of f_n
for different values of Q

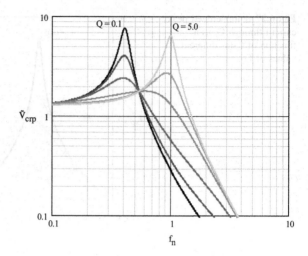

However,

$$I_{xp} = \frac{V_1}{Z_{r1}} \overline{I_{xp}} \qquad (6.123)$$

Then:

$$\overline{V_{crp}} = \frac{\overline{I_{xp}}}{Z_{r1}\omega_s C_r} \qquad (6.124)$$

We know that

$$Z_{r1}\omega_s C_r = \frac{\omega_s}{\omega_{r1}} \qquad (6.125)$$

Which implies

$$Z_{r1}\omega_s C_r = f_n \qquad (6.126)$$

Therefore:

$$\overline{V_{crp}}(\lambda, f_n, Q) = \frac{\overline{I_{xp}}(\lambda, f_n, Q)}{f_n} \qquad (6.127)$$

Figure 6.28 shows the curves generated by Eq. (6.127), where the normalized
value of the voltage across the resonant capacitor is represented as a function of the
normalized resonant frequency, for different values of the quality factor.

6.9.1   *Numerical Example 5*

A LLC converter operates with the following parameters:

- $V_1 = 400$ V
- $L_r = 14$ µH
- $L_m = 70$ µH
- $C_r = 47$ nF
- $f_n = 1$
- $\lambda = 0.2$
- $a = 1$
- $P_o = 2000$ W

Find the input impedance, current, and voltage across the resonant capacitor at the operation point.

Solution

(a) Calculation of the parameterized input impedance.

$$R_o = \frac{V_o^2}{P_o} = \frac{400^2}{2000} = 80\,\Omega$$

$$R_{ac} = \frac{8}{\pi^2} R_o = \frac{8}{\pi^2} \times 80 = 64.85\,\Omega$$

$$Q = \frac{\sqrt{\frac{L_r}{C_r}}}{R_{ac}} = \frac{\sqrt{\frac{14\times10^{-6}}{47\times10^{-9}}}}{64.85} = 0.266$$

$$|Z_1| = \sqrt{\left(\frac{Qf_n^2}{Q^2f_n^2+\lambda^2}\right)^2 + \left(\frac{f_n^2-1}{f_n} + \frac{\lambda f_n}{Q^2f_n^2+\lambda^2}\right)^2} = 3.004$$

$$Z_{r1} = \sqrt{\frac{L_r}{C_r}} = \sqrt{\frac{14\times10^{-6}}{47\times10^{-9}}} = 17.26\,\Omega$$

$$\boxed{|Z_1| = Z_{r1}|\overline{Z_1}| = 17.26 \times 3.004 = 51.84\,\Omega}$$

(b) Calculation of the peak value of the current in the resonant circuit.

$$\boxed{I_{xp} = \frac{V_{xp}}{|Z_1|} = \frac{4V_1}{\pi|Z_1|} = \frac{4\times400}{\pi\times51.84} = 9.82\,A}$$

(c) Calculation of the peak voltage across the resonant capacitor.

$$\boxed{V_{cp} = \frac{I_{xp}}{2\pi f_s C_r} = 169.6\,V}$$

The reader is encouraged to verify these results through numerical simulation.

We should know that these results are obtained considering the first harmonic approximation converter equivalent circuit and that the simulation results may be slightly different.

6.10 Design Methodology and Example

This numerical example provides a simplified design of a LLC converter, with the following specifications:

- $P_o = 250$ W
- $V'_o = 400$ V
- $V_{1max} = 400$ V
- $V_{1min} = 340$ V
- $\lambda = 0.2$
- $f_{max} = 100$ kHz

Simplifying, a transformer with unit turns ratio and ideal components will be considered. To simplify the design method description, it will be divided in steps.

(a) First Step

Let's adopt $f_{nmax} = 1$ In this way, the converter will operate in discontinuous conduction mode and the switching losses in the rectifier diodes of the output stage will be theoretically be zero.

Therefore,

$$f_{r1} = 100\,kHz \tag{6.128}$$

Since

$$f_{r1} = \frac{1}{2\pi\sqrt{C_r L_r}} \tag{6.129}$$

it is obtained (6.130).

$$C_r L_r = \frac{1}{4\pi^2 10^{10}} \tag{6.130}$$

Thus, the first relation between L_r and C_r is obtained.

(b) Second Step

The minimum static gain is given by (6.131).

$$M_{min} = \frac{V'_o}{V_{1max}} = 1 \tag{6.131}$$

The maximum static gain is given by (6.132).

$$M_{max} = \frac{V'_o}{V_{1min}} = 1.176 \tag{6.132}$$

(c) Third Step

Let's remind the expression of the parameterized static gain (6.133), previously deduced.

$$M = \frac{f_n^2}{\sqrt{\left[f_n^2(\lambda + 1) - \lambda\right]^2 + \left[f_n Q\left(1 - f_n^2\right)\right]^2}} \tag{6.133}$$

In the considered operation region, the static gain is maximum when the converter has no load, which means $Q = 0$.

In this condition, the converter must operate with its minimum frequency. Then, from (6.133), it is obtained:

$$M_{max} = \frac{f_{nmin}^2}{f_{nmin}^2(\lambda + 1) - \lambda} \tag{6.134}$$

Then:

$$f_{nmin} = \sqrt{\frac{\lambda M_{max}}{(\lambda + 1)M_{max} - 1}} \tag{6.135}$$

With the values of $\lambda = 0.2$ and $M_{max} = 1.176$, we obtain:

$$f_{nmin} = 0.756 \tag{6.136}$$

$$f_{min} = f_{nmin} f_{max} = 75.6\,kHz \tag{6.137}$$

In Fig. 6.29, the static gain curves as a function of Q are shown for the extreme values of f_n, which are $f_{nmin} = 0.756$ and $f_{nmax} = 1$.

(d) Fourth Step

In this design step, the value of Q_{max} is chosen. With the relation (6.138), (6.140) can be obtained.

Fig. 6.29 Static gain of the converter as a function of Q for values of f_n between 0.756 and 1

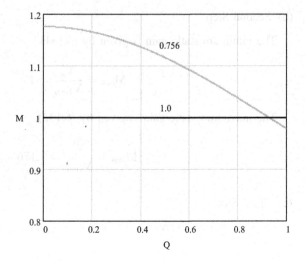

$$Q_{max} = \frac{1}{R_{ac}} \sqrt{\frac{L_r}{C_r}} \tag{6.138}$$

$$\frac{L_r}{C_r} = (Q_{max} R_{ac})^2 \tag{6.139}$$

$$R_{ac} = \left(\frac{4V'_o}{\pi}\right)^2 \frac{1}{2P_o} = 518.76\,\Omega \tag{6.140}$$

R_{ac} represents the load resistance reflected to the primary side of the transformer. The criteria to choose Q_{max} must be adopted. It has been shown that the peak current in the resonant capacitor C_r is given by (6.141).

$$I_{1p}(Q) = \frac{4V_{1max}}{\pi} \frac{1}{|Z_1(Q)|} \tag{6.141}$$

where:

$$|Z_1(Q)| = Z_r(Q) \sqrt{\left(\frac{Qf_n^2}{Q^2 f_n^2 + \lambda^2}\right)^2 + \left(\frac{f_n^2 - 1}{f_n} + \frac{\lambda f_n}{Q^2 f_n^2 + \lambda^2}\right)^2} \tag{6.142}$$

According to expressions (6.130) and (6.139), the parameters C_r, L_r, L_m, Z_1 and consequently the current value in the resonant circuit depend on Q_{max}. For the parameters previously obtained in this design example, the curve of the peak current value as a function of Q_{max} shown in Fig. 6.30 is generated.

Fig. 6.30 Peak value of the resonant tank current as a function of Q_{max}

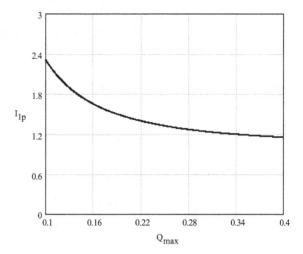

Note that an increment of Q_{max} causes a reduction of I_{1p}. However, this causes an increment in the magnetizing inductance L_m and decreases the peak value of the magnetizing current, given by (6.143).

$$I_{mp} = V'_o \tan\left(\frac{\pi}{2f_n}\sqrt{\frac{\lambda}{\lambda+1}}\right)\sqrt{\frac{C_r}{L_r + L_m}} \qquad (6.143)$$

For this design example, the value of I_{mp} as a function of Q_{max} is represented in Fig. 6.31.

Fig. 6.31 Peak value of the magnetizing current as a function of Q_{max}

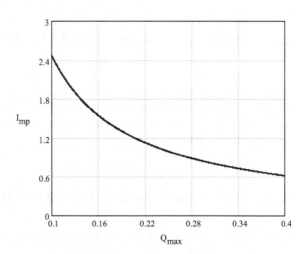

The value of I_{mp} must be larger than I_{mpmin}, to ensure soft switching with no load. Let's adopt $I_{mpmin} = 1$ A. Therefore, $Q_{max} = 0.237$. The values of L_r and C_r can be found by employing (6.144) and (6.145).

$$C_r L_r = \frac{1}{4\pi^2 f_{max}^2} \tag{6.144}$$

$$\frac{L_r}{C_r} = (Q_{max} R_{ac})^2 \tag{6.145}$$

With the calculated parameters

$$f_{max} = 100\,\text{kHz} \tag{6.146}$$

$$R_{ac} = 518.76\,\Omega \tag{6.147}$$

$$Q_{max} = 0.237 \tag{6.148}$$

It is obtained:

$$C_r = 12.94\,\text{nF} \tag{6.149}$$

$$L_r = 195.7\,\mu\text{H} \tag{6.150}$$

$$L_m = 978.4\,\mu\text{H} \tag{6.151}$$

The reader is invited and encouraged to simulate the designed LLC converter to verify the results.

6.11   Detailed Description of the LLC's Operation Modes

The previous analysis considered the first harmonic approximation equivalent circuit of the LLC converter, where all voltages and currents were sinusoidal.

In this section, the topological states and the main waveforms of the LLC converter for frequencies lower, equal or larger than the resonant frequency will be described, for rectangular voltage waveforms.

6.11.1   Operation at the Resonance Frequency ($f_s = f_{r1}$)

The relevant waveforms for $f_s = f_{r1}$ are shown in Fig. 6.32.

Fig. 6.32 Relevant waveforms at resonance

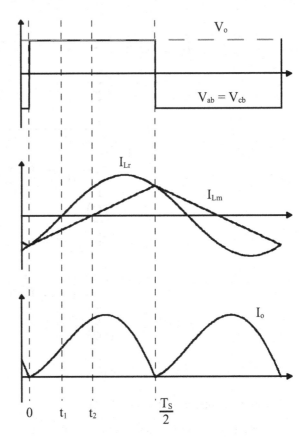

For this operating mode, the current i_o in the load is at the boundary between the discontinuous and continuous mode and the voltage at the terminals of L_m is always equal to $\pm V_o$, or V_{ab}.

In the interval $[0, T_s/2]$, three topological states occur, represented in Fig. 6.33, for the time intervals $[0, t_1]$, $[t_1, t_2]$ and $[t_2, T_s/2]$, respectively.

All components are considered ideal and the parameters of the rectifier stage and the load are all reflected to the primary side of the transformer.

(A) First Topological State $(0, t_1)$

The voltage V_{ab} is positive. The negative currents i_{Lr} and i_{Lm} rise sinusoidally and linearly, respectively. This stage ends at the instant t_1, when $i_{Lr} = 0$.

(B) Second Topological State (t_1, t_2)

During this topological state the resonant current is positive, increasing sinusoidally, while the magnetizing current, that is negative, increases linearly.

Fig. 6.33 Topological states at resonance for the time intervals **a** [0, t₁], **b** [t₁, t₂] and **c** [t₂, T_s/2]

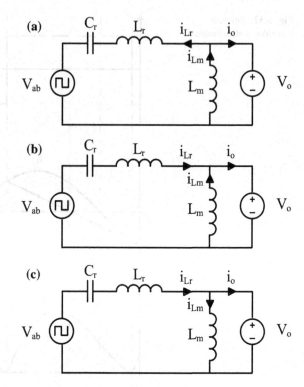

(C) Third Topological State (t₂, T_S/2)

During this topological state, both currents i_{Lr} and i_{Lm} are positive; i_{Lr} is sinusoidal while i_{Lm} continues to rise linearly.

At low loads, the current i_o may become discontinuous and more topological states may occur than those presented in Fig. 6.33.

6.11.2   Operation Below Resonant Frequency ($f_s < f_{r1}$)

The relevant waveforms for $f_s < f_{r1}$ are shown in Fig. 6.34.

In the interval [0, T_s/2], four topological states take place, represented in Fig. 6.35, for the time intervals [0, t₁], [t₁, t₂], [t₂, t₃] and [t₃, T_s/2], respectively.

(A) First Topological State (0, t₁)

The voltages V_{ab} and V_{cb} are positive and $V_o > V_1$. The currents i_{Lr} and i_{Lm} are negative. The current i_{Lr} grows sinusoidally as i_{Lm} grows linearly. This topological state ends at the instant t_1, when $i_{Lr} = 0$.

Fig. 6.34 Relevant
waveforms for $f_s < f_{r1}$

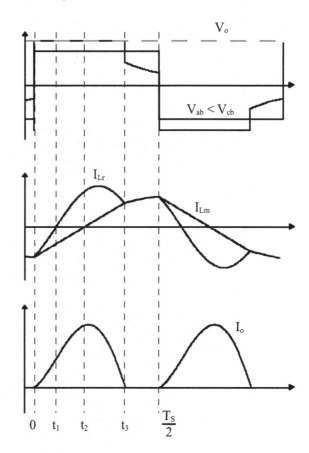

(B) Second Topological State (t_1, t_2)

The current i_{Lr} is sinusoidal and positive. It increases linearly, while i_{Lm}, negative, rises linearly. This topological state ends at the instant t_2, when $i_{Lm} = 0$.

(C) Third Topological State (t_2, t_3)

During this time interval, i_{Lr} and i_{Lm} are positive. This stage ends at time $t = t_3$, when $i_{Lr} = i_{Lm}$ and $i_o = 0$.

(D) Fourth Topological State (t_3, $T_S/2$)

During this topological state, $i_o = 0$ and $i_{Lr} = i_{Lm}$. It ends at $t = T_S/2$, when the voltage V_{ab} changes polarity and becomes negative.

The operation for $f_s < f_{r1}$ differs from the operation of $f_s = f_{r1}$ in two aspects:

- The load voltage V_o is larger than the supply voltage V_1;
- The current i_o at the output of the rectifier is discontinuous.

Fig. 6.35 Topological states at switching frequency below resonant frequency

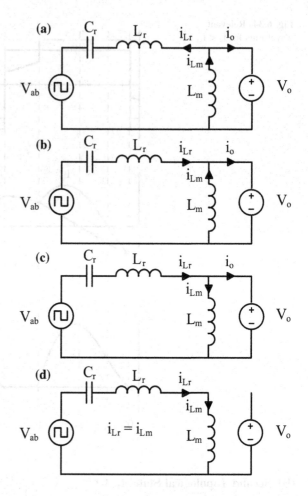

6.11.3   Operation Above Resonant Frequency ($f_s > f_{r1}$)

The relevant waveforms for $f_s > f_{r1}$ are shown in Fig. 6.36.

In the time interval $[0, T_s/2]$, four topological states take place, represented in Fig. 6.37, for the time subintervals $[0, t_1]$, $[t_1, t_2]$, $[t_2, t_3]$ and $[t_3, T_s/2]$, respectively.

(A) First Topological State $(0, t_1)$

During this time interval, V_{ab} and V_{cb} are positive. The currents i_{Lr} and i_{Lm} are negative. This interval ends at time $t = t_1$, when i_{Lr} reaches zero.

(B) Second Topological State (t_1, t_2)

During this time interval, V_{ab} and V_{cb} remain positive. The current i_{Lr} is positive while i_{Lm} remains negative and rises linearly. This topological state ends at time $t = t_2$, when the magnetizing current reaches zero.

Fig. 6.36 Relevant
waveforms at $f_s > f_{r1}$

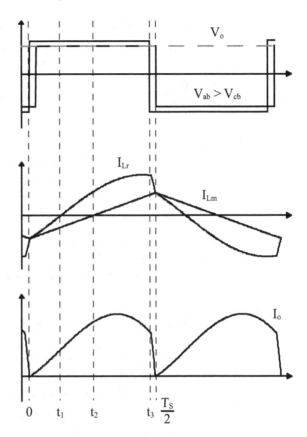

(C) **Third Topological State (t_2, t_3)**

During this time, i_{Lr} and i_{Lm} are positive. This topological state ends at time $t = t_3$, when the input voltage V_{ab} is inverted and becomes negative.

(D) **Fourth Topological State (t_3, $T_S/2$)**

During this topological state, the current i_{Lr} decreases until it becomes equal to the magnetizing current i_{Lm}, at time $t = T_S/2$.

In the operation description at $f_s > f_{r1}$, the time starts to be counted at the moment the voltage V_{cb} becomes positive, which makes the description of the converter operation more understandable. In this operation mode, the polarity reversal of the voltage V_{ab} occurs before the current i_o naturally reaches zero, which causes forced blocking of the rectifier diodes, and consequently increases the switching losses, sacrificing the efficiently. The operation for $f_s > f_{r1}$ is undesirable for these reasons, in comparison to other operation modes, and should be avoided.

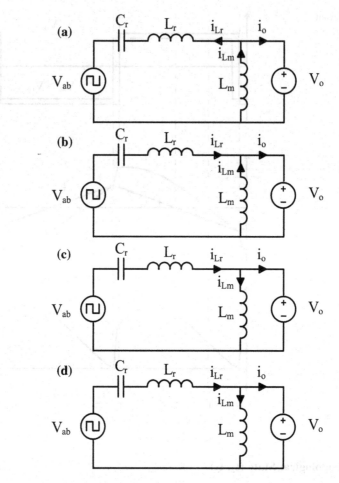

Fig. 6.37 Topological states above resonance $f_s > f_{r1}$

6.12 Summary of Properties of the LLC Converter

According to what has been studied in this chapter, the LLC converter has several advantages over the other resonant converters, to be specific:

- It allows the regulation of the load voltage for large variations of input voltage and power processed with a small variation of the switching frequency;
- With the appropriate combination of parameters, it can operate with soft switching of the active switches from full to zero load;
- As the voltages and currents are sinusoidal, it requires smaller and lower cost EMI filters to comply with EMI standards;

- All parasitic elements such as semiconductor junction capacitance, body diodes and leakage and magnetizing inductances are beneficial for the correct operation of the LLC converter and favors the soft commutation of the power switches.

6.13 Proposed Exercises

(1) The LLC converter shown in Fig. 6.1 operates with the following parameters:

- $V_1 = 100$ V
- $f_s = 100$ kHz
- $C_r = 633$ nF
- $L_r = 4$ μH
- $L_m = 20$ μH
- $a = 10$
- $R_o = 1000$ Ω

Calculate:

(a) The inductance's ratio λ;
(b) The quality factor Q;
(c) The resonant frequency f_{r1};
(d) The resonant frequency f_{r2};
(e) The normalized frequency f_n;
(f) The load voltage V_o;
(g) The power transferred to the load resistor R_o;
(h) The peak value of the magnetizing current;
(i) The peak value of the current through the inductor L_r;
(j) The average value of the current in the power supply V_1;
(k) The phase angle ϕ between V_{ab} and i_{Lr};
(l) The peak value of the voltage across the resonant capacitor C_r;
(m) The average value of the current in an output rectifier diode;
(n) The rms current value in an inverter bridge switch.

Answers:
(a) $\lambda = 0.2$; (b) Q = 0.31; (c) $f_{r1} = 100$ kHz; (d) $f_{r2} = 40,830$ kHz; (e) $f_n = 1$;
(f) $V_o = 1000$ V; (g) $P_o = 1000$ W; (h) $I_{mp} = 12.5$ A; (i) $I_p = 20.07$ A;
(j) $I_1 = 10$ A; (k) $\phi = 0.672$ rad; (l) $V_{crp} = 50.47$ V; (m) $I_{Daver} = 0.5$ A;
(n) $I_{srms} = 10.04$ A

(2) A LLC converter operates with the following parameters:

- $V_1 = 200$ V
- $f_s = 150$ kHz
- $C_r = 100$ nF
- $L_r = 7$ μH

- $L_m = 35 \ \mu H$
- $a = 0.25$
- $R_o = 500 \ \Omega$

Using the equations deduced for the first harmonic approximation, determine:

(a) The voltage across the load resistor R_o;
(b) the power delivered to the load.

Answers: (a) $V_o = 888$ V and (b) $P_o = 1578$ W.

The reader is encouraged to simulate the converter with the above parameters, to verify the error introduced by the first harmonic approximation.

(3) A LLC converter operates at normalized frequency $f_n = 1$. Demonstrate that the peak value I_p of the sinusoidal current that flows through the inductor L_r is given by expression

$$I_p = \sqrt{I_{mp}^2 + \left(\frac{\pi}{2} I_o\right)^2}$$

where I_{mp} and I_o are the value of the magnetizing current at the instant of commutation, and the average value of the load current, respectively.

(4) Prove that the angle ϕ between the sinusoidal current in the inductor L_r and the voltage at the input of the resonant circuit is given by the following expression:

$$\tan(\phi) = \frac{2}{\pi} \frac{I_{mp}}{I_o}$$

(5) In the text of the present chapter it was demonstrated that for the first harmonic approximation, $M = 1$ (static voltage gain) for $f_n = 1$ (normalized switching frequency). Prove that this result is also valid for rectangular voltages, as in a real converter.

References

1. Schmidmer, E.G.: A new high frequency resonant converter topology. In: HFPC Conference Record, pp. 9–16 (1988)
2. Sevems, R.P.: Topologies for three-element resonant converters. In: IEEE Transactions on Power Electronics, vol. 1.7, no. 1, pp. 89–98 (1992)
3. Lazar, J.F., Martinelli, R.: Steady-state analysis of the LLC series resonant converter. In: Sixteenth Annual IEEE Applied Power Electronics Conference and Exposition, APEC 2001, vol. 2, pp. 728–735 (2001)
4. Yang, B., Lee, F.C., Zhang, A.J., Huang, G.: LLC resonant converter for front end DC/DC conversion. In: Seventeenth Annual IEEE Applied Power Electronics Conference and Exposition, APEC 2002, vol. 2, pp. 1108–1112 (2002)

Chapter 7
Full-Bridge ZVS-PWM Converter with Capacitive Output Filter

Nomenclature

V_i	Input DC voltage
V_o	Output DC voltage
P_o	Output power
C_o	Output filter capacitor
R_o	Output load resistor
ZVS	Zero voltage switching
ϕ	Angle between the leading leg and the lagging leg
q	Static gain
D	Duty cycle
f_s	Switching frequency
T_s	Switching period
t_d	Dead time
n	Transformer turns ratio
V'_o	Output DC voltage referred to the transformer primary side
i_o	Output current
i'_o	Output current referred to the transformer primary side
i'_{oC}	Output current referred to the primary in CCM
i'_{oD}	Output current referred to the primary in DCM
I'_o	$\left(\overline{I'_o}\right)$ average output current referred to the primary and its normalized value
I'_{oC}	$\left(\overline{I'_{oC}}\right)$ average output current in CCM referred to the primary and its normalized value
I'_{oD}	$\left(\overline{I'_{oD}}\right)$ average output current in DCM referred to the primary and its normalized value
$I'_{oL}\left(\overline{I'_{oL}}\right)$	Average output current in the limit (critical conduction mode) between CCM/DCM, referred to the primary winding and its normalized value
S_1 and S_2	Switches in the leading leg
S_3 and S_4	Switches in the lagging leg

© Springer International Publishing AG, part of Springer Nature 2019
I. Barbi and F. Pöttker, *Soft Commutation Isolated DC-DC Converters*,
Power Systems, https://doi.org/10.1007/978-3-319-96178-1_7

v_{g1}, v_{g2}, v_{g3} and v_{g4}	Switches S_1, S_2, S_3 and S_4 drive signals, respectively
D_1, D_2, D_3 and D_4	Diodes in anti-parallel to the switches (MOSFET—intrinsic diodes)
C_1, C_2, C_3 and C_4	Capacitors in parallel to the switches (MOSFET—intrinsic capacitors)
L_c	Transformer leakage inductance or an additional inductor
i_{Lc}	Inductor current
I_{Lc}	$(\overline{I_{Lc}})$ inductor peak current and its normalized value, commutation current for S_1 and S_3
I_{Lc_C}	$(\overline{I_{Lc\_C}})$ inductor peak current in CCM and its normalized value
I_{Lc_D}	$(\overline{I_{Lc\_D}})$ inductor peak current in DCM and its normalized value
$\overline{I_{Lc\,RMS}}$	Inductor L_c normalized RMS current
$I_{Lc\,RMS\_C}$	$(\overline{I_{Lc\,RMS\_C}})$ inductor L_c RMS current in CCM and its normalized value
$I_{Lc\,RMS\_D}$	$(\overline{I_{Lc\,RMS\_D}})$ inductor L_c RMS current in DCM and its normalized value
I_1	$(\overline{I_1})$ inductor current at the end of the first and fourth step of operation (commutation current for S_2 and S_4) and its normalized value
v_{ab}	Full bridge ac voltage, between points "a" and "b"
v_{cb}	Inductor voltage, between points "c" and "b"
v_{ac}	Voltage at the ac side of the rectifier, between points "a" and "c"
v_{S1}, v_{S2}, v_{S3} and v_{S4}	Voltage across switches
i_{S1}, i_{S2}, i_{S3} and i_{S4}	Current in the switches
i_{C1}, i_{C2}, i_{C3} and i_{C4}	Current in the capacitors
ΔT	Time interval which $v_{ab} = \pm V_i$
Δt_1	Time interval of the first step of operation ($t_1 - t_0$)
Δt_2	Time interval of the second step of operation ($t_2 - t_1$)
Δt_3	Time interval of the third step of operation ($t_3 - t_2$)
Δt_4	Time interval of the fourth step of operation ($t_4 - t_3$)
Δt_5	Time interval of the fifth step of operation ($t_5 - t_4$)
Δt_6	Time interval of the sixth step of operation ($t_6 - t_5$)
A_1, A_2 and A_3	Areas

7.1 Introduction

In this chapter, the Full-Bridge Zero-Voltage-Switching Pulse-Width-Modulation Converter (FB-ZVS-PWM) with a Capacitive Output Filter is described [1]. Such converter is suitable for high-power and high-voltage applications, such as industrial, commercial or residential battery chargers as well as embedded battery chargers for hybrid or electrical vehicles. The chapter is organized to present at first the topology of the converter and its basic operation with ideal components, followed by a detailed analysis of the operation of the converter in Continuous and Discontinuous Conduction Modes as well as the Zero Voltage Switching operation and a design example.

The schematics of the FB-ZVS-PWM Converter with capacitive output filter, shown in Fig. 7.1, consists of a full bridge inverter, an inductor, a transformer, a full bridge diode rectifier and a capacitive output filter. The diodes (D_1, D_2, D_3 and D_4) in anti-parallel to the switches (S_1, S_2, S_3 and S_4) and the parallel capacitors (C_1, C_2, C_3 and C_4) may be the parasitic components of a MOSFET or additional components in parallel to the switches, as in case of IGBTs. The inductance L_c may be the transformer leakage inductance or an additional inductor, when necessary. This inductor is important to the power transfer to the load as well as to achieve soft commutation. All the operation stages are linear, simplifying the analysis and the design.

This converter operates with a fixed switching frequency and the voltages across the switches are limited to the DC link voltage (V_i). To achieve zero voltage switching (ZVS) the two legs operate with a phase-shift modulation and the switches of each leg are gated complementary.

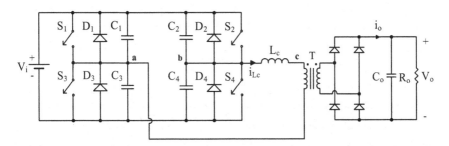

Fig. 7.1 Full-bridge ZVS-PWM converter, with capacitive output filter

7.2   Circuit Operation

In this section, the converter shown in Fig. 7.2, without the capacitors in parallel to the switches is analyzed (the soft commutation analysis is presented in Sect. 7.4). The following assumptions are made:

- all components are considered ideal;
- the converter is on steady-state operation;
- the output filter is represented as a DC voltage V'_o, whose value is the output voltage referred to the primary winding of the transformer;
- current that flows through the switches is unidirectional, i.e. the switches allow the current to flow only in the direction of the arrow;
- the diodes in anti-parallel to the switches conduct separately from the switches, so the analysis is not limited to a MOSFET.

The switches' drive signals with phase shift modulation are shown in Fig. 7.3 for both the leading leg (S_1 and S_3) and the lagging leg (S_2 and S_4) as well as the voltage v_{ab}. No dead time is considered at this point.

Fig. 7.2 Ideal equivalent circuit Full-Bridge ZVS-PWM converter, with output capacitive filter

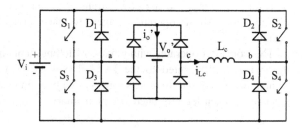

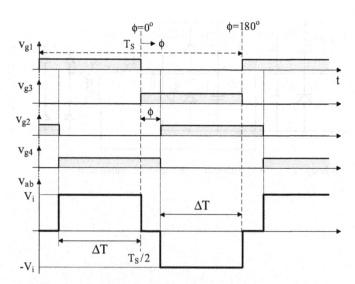

Fig. 7.3 Phase shift modulation—basic operation of FB ZVS-PWM

In the phase shift modulation, the switching frequency is fixed and the switches duty cycle are 50%. The lagging leg drive signals are shifted of an angle ϕ in relation to the leading leg drive signals, allowing the power transfer to be controlled. If $\phi = 0°$, v_{ab} has its maximum RMS value since switches S_1 and S_4 always conduct at the same time ($v_{ab} = V_i$), and switches S_2 and S_3 also conduct at the same time ($v_{ab} = -V_i$). When ϕ increases, the v_{ab} RMS value decreases because switches S_1 and S_2 as well as S_3 and S_4 conduct at the same time resulting in $v_{ab} = 0$. The ϕ angle varies from 0° (maximum power) to 180° (zero power).

7.2.1 Continuous Conduction Mode

The operation of the converter is described during one switching period that is divided into six time intervals, each one representing a different state of the switches. In CCM, the rectifier diodes conduct the inductor current in the six time intervals.

(A) Time Interval Δt_1 ($t_0 \leq t \leq t_1$)

The first time interval (shown in Fig. 7.4) starts at $t = t_0$, when switch S_1 is gated on and switch S_3 is turned off. Although switch S_1 is gated on, the current does not flows through it because i_{Lc} is negative ($i_{Lc}(t_0) = -I_{Lc\_c}$). Hence diode D_1 conducts the current together with switch S_2 that was conducting prior to this time interval. The voltage v_{ab} is zero and the inductor voltage v_{cb} is V_o', leading the inductor to deliver power to the load.

Fig. 7.4 Topological state in CCM for time interval Δt_1

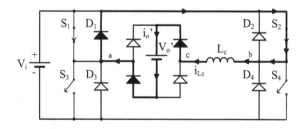

Fig. 7.5 Topological state in CCM for time interval Δt_2

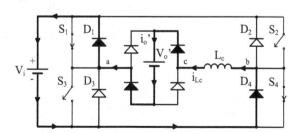

(B) Time Interval Δt_2 ($t_1 \leq t \leq t_2$)

This time interval, shown in Fig. 7.5, starts at $t = t_1$, when S_4 is gated on and S_2 is turned off. As the inductor current is still negative ($i_{Lc}(t_1) = -I_1$), the current flows through diodes D_1 and D_4, $v_{ab} = V_i$ and the inductor voltage $v_{cb} = V_i + V'_o$. So, the inductor delivers power to the load and to the DC link (V_i).

(C) Time Interval Δt_3 ($t_2 \leq t \leq t_3$)

Figure 7.6 shows the topological state for time interval Δt_3. It begins at $t = t_2$ when i_{Lc} reaches zero ($i_{Lc}(t_2) = 0$), turning off diodes D_1 and D_4 and enabling switches S_1 and S_4 to conduct, as the inductor current evolves positively. The voltage v_{ab} is V_i and the inductor voltage v_{cb} is $V_i - V'_o$, leading the DC link to deliver energy to the load and the inductor L_c.

(D) Time Interval Δt_4 ($t_3 \leq t \leq t_4$)

This time interval starts at $t = t_3$, when switch S_1 is turned off and switch S_3 is gated on. The current does not flow through switch S_3 because of its unidirectional current flow assumption. The inductor current flows through diode D_3 and switch

Fig. 7.6 Topological state in CCM for time interval Δt_3

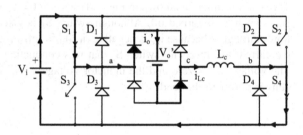

Fig. 7.7 Topological state in CCM for time interval Δt_4

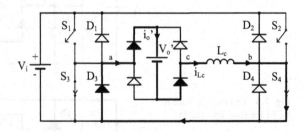

Fig. 7.8 Topological state in CCM for time interval Δt_5

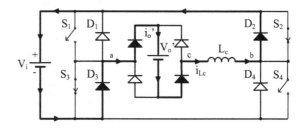

Fig. 7.9 Topological state in CCM for time interval Δt_6

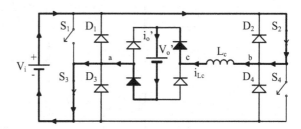

S_4, that was already conducting, as shown in Fig. 7.7. The voltage v_{ab} is zero and the inductor voltage v_{cb} is $-V'_o$, leading the inductor to deliver power to the load.

(E) Time Interval Δt_5 ($t_4 \leq t \leq t_5$)

This time interval, shown in Fig. 7.8, begins at $t = t_4$, when S_4 is turned off and S_2 is gated on. As the inductor current is still positive ($i_{Lc}(t_4) = +I_1$), the current flows through diodes D_2 and D_3, $v_{ab} = -V_i$, and the inductor voltage $v_{cb} = -V_i - V'_o$, leading the inductor to deliver power to the load and to the DC link (V_i).

(F) Time Interval Δt_6 ($t_5 \leq t \leq t_6$)

Time interval Δt_6 is shown in Fig. 7.9. It begins at $t = t_5$ when i_{Lc} reaches zero ($i_{Lc}(t_5) = 0$), turning off diodes D_2 and D_3 and enabling switches S_2 and S_3 to conduct, as the inductor current evolves negatively. The voltage v_{ab} is $-V_i$ and the inductor voltage v_{cb} is $-V_i + V'_o$, leading the DC link to deliver energy to the load and the inductor L_c. The main waveforms for the continuous conduction mode are shown in Fig. 7.10.

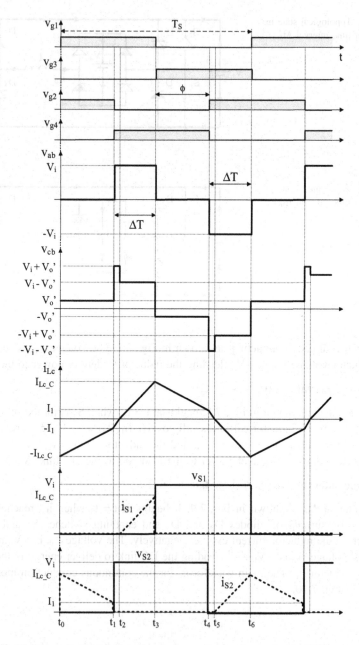

Fig. 7.10 Main waveforms and time diagram for one cycle of operation in CCM

7.2.2 Discontinuous Conduction Mode

The converter is described during one switching period that is divided into six time intervals, each one representing a different state of the switches. In DCM the rectifier diodes conduct the inductor current only in four time intervals, because the inductor current remains in zero in the time intervals Δt_2 and Δt_5.

(A) Time Interval Δt_1 ($t_0 \leq t \leq t_1$)

Time Interval Δt_1 (shown in Fig. 7.11) starts at $t = t_0$, when switch S_1 is gated on and switch S_3 is turned off. Although the switch S_1 is gated on, the current does not flow through it because i_{Lc} is negative ($i_{Lc}(t_0) = -I_{Lc\_D}$), so diode D_1 conducts the current together with switch S_2 that was conducting prior to this interval. The voltage v_{ab} is zero and the inductor voltage v_{cb} is V_o', leading the inductor to deliver power to the load.

(B) Time Interval Δt_2 ($t_1 \leq t \leq t_2$)

This time interval starts at $t = t_1$, when the inductor current reaches zero ($i_{Lc}(t_1) = 0$), turning off D_1 and S_2. No current flows through the circuit as shown in Fig. 7.12, and $v_{ab} = 0$ and $v_{cb} = 0$.

Fig. 7.11 Topological state in DCM for time interval Δt_1

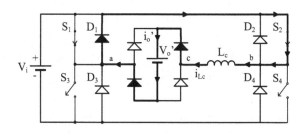

Fig. 7.12 Topological state in DCM for time interval Δt_2

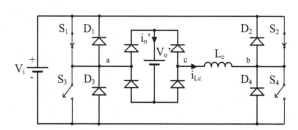

Fig. 7.13 Topological state in DCM for time interval Δt_3

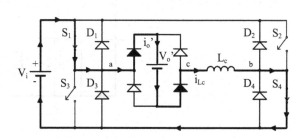

Fig. 7.14 Topological state
in DCM for time interval Δt_4

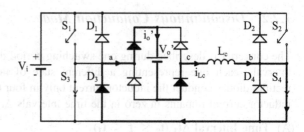

(C) Time Interval Δt_3 ($t_2 \leq t \leq t_3$)

This time interval, shown in Fig. 7.13, starts at t = t_2, when S_4 is gated on and S_2 is turned off. As the current is positive, it flows through switches S_1 and S_4. During this interval $v_{ab} = V_i$, the inductor voltage $v_{cb} = V_i - V'_o$, and the DC link deliver power to the inductor and to the load.

(D) Time Interval Δt_4 ($t_3 \leq t \leq t_4$)

This time interval starts at t = t_3 when switch S_1 is turned off and switch S_3 is gated on. The current does not flow through switch S_3 because of its unidirectional current flow assumption. The inductor current flows through diode D_3 and switch S_4 that was already conducting, as show in Fig. 7.14. The voltage v_{ab} is zero and the inductor voltage v_{cb} is $-V'_o$, leading the inductor to deliver power to the load.

(E) Time Interval Δt_5 ($t_4 \leq t \leq t_5$)

Time interval Δt_5 starts at t = t_4, when the inductor current reaches zero ($i_{Lc}(t_1) = 0$). No current flows through the circuit as shown in Fig. 7.15. During this time interval $v_{ab} = 0$ and $v_{cb} = 0$.

Fig. 7.15 Topological state
in DCM for time interval Δt_5

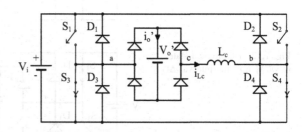

Fig. 7.16 Topological state
in DCM for time interval Δt_6

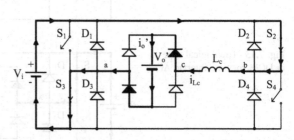

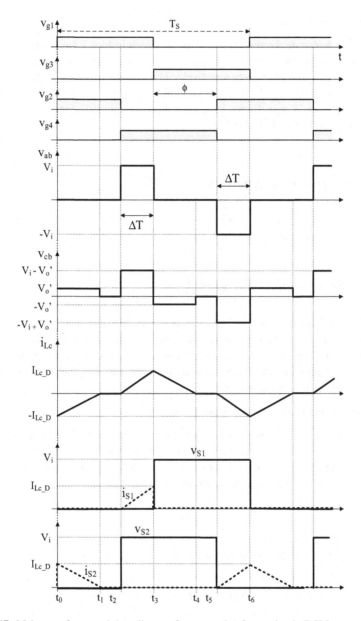

Fig. 7.17 Main waveforms and time diagram for one cycle of operation in DCM

(F) **Time Interval Δt_6 ($t_5 \leq t \leq t_6$)**

The topological state for time interval Δt_6 is shown in Fig. 7.16. It begins at $t = t_5$ when S_2 is turned off and S_4 is gated on. The inductor current flows through switches S_2 and S_3 and evolves negatively. The voltage v_{ab} is $-V_i$ and the inductor voltage v_{cb} is $-V_i + V'_o$, leading the DC link to deliver energy to the load and the inductor. The main waveforms indicating the time intervals for thediscontinuous conduction mode are shown in Fig. 7.17.

7.3   Mathematical Analysis

In this section the FB-ZVS-PWM commutation currents (I_1 and I_{Lc}), the output characteristics and the RMS inductor current are presented.

7.3.1   Commutation Currents

The commutation currents for the leading leg (I_{Lc}) and the lagging leg (I_1) are obtained analyzing the equations for the different intervals of operation for CCM and DCM. The commutation intervals are not considered because these intervals are very fast compared to the other ones, so the inductor current can be considered constant during the commutation.

(A) **Continuous Conduction Mode (CCM)**

In CCM S_1 and S_3 commutate with the inductor peak current I_{Lc_C} and S_2 and S_4 commutate with a smaller current I_1, so this commutation is more critical.

Analyzing the second time interval ($t_1 \leq t \leq t_2$) in CCM (Sect. 7.2.1; Fig. 7.5) the following equation is written:

$$\left(V_i + V'_o\right) = L_c \frac{di_{Lc}(t)}{dt} \tag{7.1}$$

Integrating (7.1) from t_1 to t_2, (7.2) is obtained.

$$\left(V_i + V'_o\right) \int_{t_1}^{t_2} dt = L_c \int_{-I_1}^{0} dt \tag{7.2}$$

Solving the integral and isolating I_1:

$$I_1 = \frac{V_i + V'_o}{L_k} \Delta t_2 \tag{7.3}$$

In the third interval of operation ($t_2 \leq t \leq t_3$) in CCM (Sect. 7.2.1; Fig. 7.6) the following equation derives:

$$V_i - V'_o = L_c \frac{di_{Lc}(t)}{dt} \tag{7.4}$$

Integrating (7.3) from t_2 to t_3, (7.5) is obtained.

$$\left(V_i - V'_o\right) \int_{t_2}^{t_3} dt = L_c \int_0^{I_{Lc\_C}} dt \tag{7.5}$$

Solving the integral and isolating the inductor peak current in CCM (I_{Lc_C}), (7.6) is obtained.

$$I_{Lc\_C} = \frac{V_i - V'_o}{L_c} \Delta t_3 \tag{7.6}$$

In the fourth interval of operation ($t_3 \leq t \leq t_4$) in CCM (Sect. 7.2.1; Fig. 7.7) the following equation is written:

$$-V'_o = L_c \frac{di_{Lc}(t)}{dt} \tag{7.7}$$

Integrating (7.7) from t3 to t4:

$$\int_{t_3}^{t_4} V'_o dt = -L_c \int_{I_{Lc\_C}}^{I_1} di \tag{7.8}$$

Solving the integral and isolating I_1:

$$I_1 = I_{Lc\_C} - \frac{V'_o}{L_c} \Delta t_4 \tag{7.9}$$

As the FB-ZVS-PWM is symmetric in one switching period, the following time intervals are equal: $\Delta t_1 = \Delta t_4, \Delta t_2 = \Delta t_5, \Delta t_3 = \Delta t_6$.

The time interval which $v_{ab} = \pm V_i$ is defined as ΔT, as shown in Fig. 7.10, so (7.10) and (7.11) may be written as:

$$\Delta T = \Delta t_3 + \Delta t_2 \tag{7.10}$$

$$\frac{T_s}{2} = \Delta T + \Delta t_4 \tag{7.11}$$

Isolating Δt_4 gives:

$$\Delta t_4 = \frac{T_s}{2} - \Delta T \tag{7.12}$$

The static gain and the duty cycle are calculated as follows:

$$q = \frac{V'_o}{V_i} \tag{7.13}$$

$$D = \frac{2\Delta T}{T_s} \tag{7.14}$$

Substituting (7.3), (7.6) and (7.12) in (7.9):

$$\left(\frac{V_i + V'_o}{L_c}\right)\Delta t_2 = \left(\frac{V_i - V'_o}{L_c}\right)\Delta t_3 - \frac{V'_o}{L_c}\left(\frac{T_s}{2} - \Delta T\right) \tag{7.15}$$

Substituting (7.10) in (7.15), Δt_3 is obtained.

$$\Delta t_3 = \frac{\Delta T}{2} + \frac{V'_o}{V_i}\frac{T_s}{4} \tag{7.16}$$

Substituting (7.13) and (7.14) in (7.16) the normalized third interval of operation is calculated as follows:

$$\overline{\Delta t_3} = \frac{\Delta t_3}{T_s} = \frac{D + q}{4} \tag{7.17}$$

Substituting (7.17) and (7.14) in (7.10), the normalized fifth interval of operation is obtained.

$$\overline{\Delta t_5} = \frac{\Delta t_5}{T_s} = \frac{D - q}{4} \tag{7.18}$$

Substituting (7.17) in (7.6), the inductor peak current in CCM (I_{Lc_C}) is obtained in (7.19).

$$I_{Lc\_C} = \frac{V_i - V'_o}{L_c}\frac{D + q}{4}T_s \tag{7.19}$$

Normalizing (7.19):

$$\overline{I_{Lc\_C}} = \frac{I_{Lc\_C} 4 f_s L_c}{V_i} = (1 - q)(D + q) \tag{7.20}$$

Substituting (7.18) in (7.3), the current I_1 is obtained in (7.21).

$$I_1 = T_s \frac{V_i + V_o' \frac{D - q}{4}}{L_c} \qquad (7.21)$$

Normalizing (7.21):

$$\overline{I_1} = \frac{I_1 4 f_s L_c}{V_i} = (1 + q)(D - q) \qquad (7.22)$$

(B) Discontinuous Conduction Mode (DCM)

In DCM S_1 and S_3 commutate with the inductor peak current I_{Lc_D} and when S_2 and S_4 commutate the inductor current is zero, so there is no energy to charge/discharge capacitors C_2 and C_4, leading to a dissipative commutation in this leg.

Analyzing the third time interval ($t_2 \leq t \leq t_3$) in DCM (Sect. 7.2.2; Fig. 7.13) the following equation is written:

$$V_i - V_o' = L_c \frac{di_{Lc}(t)}{dt} = \frac{L_c I_{Lc\_D}}{\Delta T} \qquad (7.23)$$

Substituting (7.14) in (7.23):

$$I_{Lc\_D} = \frac{(V_i - V_o') D T_s}{L_c \; 2} \qquad (7.24)$$

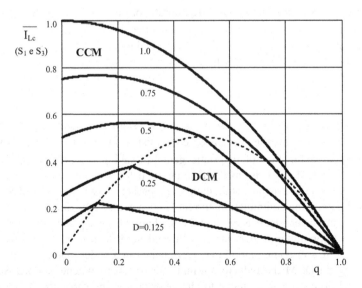

Fig. 7.18 Normalized commutation current in switches S_1 and S_3, as a function of the static gain (q), taking the duty cycle (D) as a parameter, in CCM and DCM

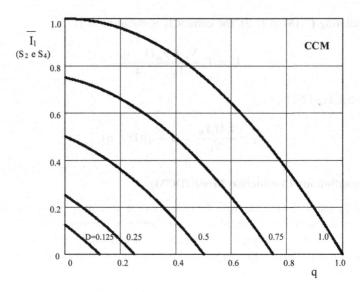

Fig. 7.19 Normalized commutation current in switches S2 and S4, as a function of the static gain (q), taking the duty cycle (D) as a parameter, in CCM

Normalizing (7.24):

$$\overline{I_{Lc\_D}} = \frac{I_{Lc\_D} 4 f_s L_c}{V_i} = 2(1-q)D \qquad (7.25)$$

Figure 7.18 presents the normalized commutation current $(\overline{I_{Lc}})$ for the switches S_1 and S_3, as a function of the static gain (q), taking the duty cycle (D) as a parameter, in CCM and in DCM. For a given static gain, as the duty cycle decreases (lower output power), the commutation current I_{Lc} also decreases, increasing the time to charge/discharge the capacitors C_1 and C_3, and soft commutation may not be achieved in all load range. If a low static gain is chosen (proper transformer turns ratio) soft commutation may be obtained for a wider load range in the leading leg, however, the switches conduction losses increase because a low static gain means higher currents.

Figure 7.19 presents the normalized commutation current $(\overline{I_1})$ for the switches S_2 and S_4, as a function of the static gain (q), taking the duty cycle (D) as a parameter, in CCM. As I_1 is smaller than the inductor peak current I_{Lc}, switches S_2 and S_4 have a more critical commutation comparing to switches S_1 and S_3. For a given static gain, the commutation current I_1 decreases, as the duty cycle decreases (lower output power), increasing the time to charge/discharge the capacitor C_2 and C_4, and soft commutation may not be achieved in all load range. If a lower static gain is chosen, soft commutation may be obtained in a wider load range, as for the leading leg. In DCM the inductor current is zero when switches S_2 and S_4 commutate and no energy is available to charge/discharge the capacitors, leading to a dissipative commutation.

7.3.2   Output Characteristics

The FB-ZVS-PWM output characteristics are obtained in this section, for continuous and discontinuous conduction mode. The commutation time intervals are disregarded.

(A) Continuous Conduction Mode (CCM)

The output current referred to the primary winding of the transformer in CCM is shown in Fig. 7.20.

The normalized area $\overline{A_1}$ is calculated as follows:

$$\overline{A_1} = \frac{2}{T_s} \frac{I_{Lc\_C} \, \Delta t_3}{2} \tag{7.26}$$

where $\Delta t_3 = t_3 - t_2$.

Substituting (7.17) and (7.20) in (7.26) gives:

$$\overline{A_1} = \frac{V_i}{16 L_c f_s} (1 - q)(D + q)^2 \tag{7.27}$$

The normalized area $\overline{A_2}$ is given by

$$\overline{A_2} = \frac{2}{T_s} \frac{(I_1 + I_{Lc\_C})\Delta t_4}{2} \tag{7.28}$$

Substituting (7.14) in (7.12), the fourth time interval is calculated:

$$\Delta t_4 = \frac{T_s}{2} - \Delta T = \frac{T_s}{2}(1 - D) \tag{7.29}$$

Substitution of (7.19), (7.21) and (7.29) into (7.28), $\overline{A_2}$ lead to Eq. (7.30).

$$\overline{A_2} = \frac{V_i}{8 f_s L_c} (1 - D)[(1 + q)(D - q) + (1 - q)(D + q)] \tag{7.30}$$

The normalized area $\overline{A_3}$ is calculated according to (7.31).

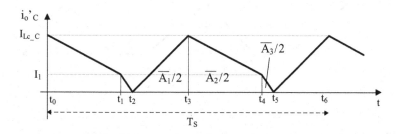

Fig. 7.20 Load current in CCM referred to the transformer primary winding

$$\overline{A_3} = \frac{2}{T_s} \frac{I_1 \Delta t_5}{2} \tag{7.31}$$

Substituting (7.22) and (7.18) in (7.31), gives:

$$\overline{A_3} = \frac{V_i}{16 f_s L_c} (1+q)(D-q)^2 \tag{7.32}$$

The average load current is the sum of the normalized areas $\overline{A_1}$, $\overline{A_2}$ and $\overline{A_3}$, which is represented by Eq. (7.33).

$$I'_{o\,C} = \frac{V_i}{8} \frac{1}{f_s L_c} (2D - D^2 - q^2) \tag{7.33}$$

Normalizing (7.33) yields the result:

$$\overline{I'_{o\,C}} = \frac{I'_{o\,C} 4 f_s L_c}{V_i} = \frac{D(2-D) - q^2}{2} \tag{7.34}$$

Equation (7.34) gives the normalized load current, referred to the transformer primary winding, as a function of the duty-cycle D and the static gain q, for continuous conduction mode (CCM).

(B) Discontinuous Current Mode (DCM)

The output current referred to the primary of the transformer in DCM is shown in Fig. 7.21.

Equation (7.35) is written for the fourth time interval:

$$V'_o = -L_r \frac{di_{Lc}(t)}{dt} = -L_c \frac{-I_{Lc\_D}}{\Delta t_4} \tag{7.35}$$

Isolating the inductor peak current in DCM gives:

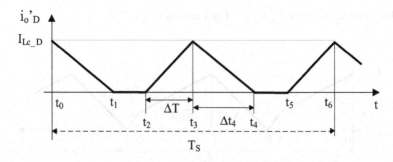

Fig. 7.21 Load current referred to the transformer primary winding, in DCM

$$I_{Lc\_D} = \frac{V_o' \, \Delta t_4}{L_c} \tag{7.36}$$

Making (7.36) equals to (7.24) yields:

$$I_{Lc\_D} = \frac{V_o' \, \Delta t_4}{L_c} = \frac{V_i - V_o'}{L_c} \Delta T \tag{7.37}$$

Substituting (7.14) in (7.37), the fourth time interval is obtained:

$$\Delta t_4 = \frac{1 - q}{q} \frac{DT_s}{2} \tag{7.38}$$

The average load current in DCM is calculated as follows:

$$I_{oD}' = \frac{2}{T_s} \left[\frac{I_{Lc\_D} \, \Delta T}{2} + \frac{I_{Lc\_D} \, \Delta t_4}{2} \right] \tag{7.39}$$

Substituting (7.37) and (7.38) in (7.39) gives:

$$I_{oD}' = \frac{1 - q}{q} \frac{V_i}{L_c} \frac{D^2}{4f_s} \tag{7.40}$$

Normalizing (7.40) yields:

$$\overline{I_{oD}'} = \frac{\overline{I_{oD}'} \, 4 L_c \, f_s}{V_i} = D^2 \left(\frac{1 - q}{q} \right) \tag{7.41}$$

Equation (7.41) gives the normalized load current, referred to the transformer primary winding, as a function of the duty-cycle D and the static gain q, for discontinuous conduction mode (DCM).

(C)   Critical Conduction Mode

In critical conduction mode, the time interval $\Delta t_5 = 0$. Thus,

$$\frac{\Delta t_5}{T_s} = \frac{D - q}{4} = 0 \tag{7.42}$$

From which we find Eq. (7.43) that represents the limit between CCM and DCM.

$$D = q \tag{7.43}$$

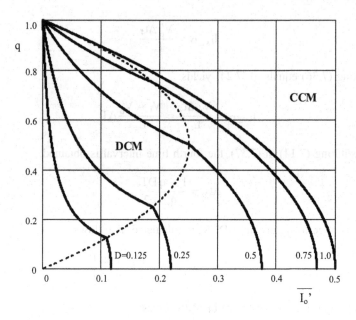

Fig. 7.22 FB-ZVS-PWM normalized output characteristics

Substituting (7.43) in (7.41) or in (7.34), (7.44) is obtained, representing the normalized average output current in critical conduction mode.

$$\overline{I'_{oL}} = \frac{\overline{I'_{oL}} \, 4 \, L_c \, f_s}{V_i} = q - q^2 \qquad (7.44)$$

The FB-ZVS-PWM output characteristics for CCM (7.36) and DCM (7.41), as well as the limit between the two conduction modes (7.44) (dashed line) are presented in Fig. 7.22. The smaller the static gain, the wider the CCM load range and the soft commutation range. However, a low static gain implies in higher currents and consequently higher conduction losses.

7.3.3 RMS Inductor Current

(A) CCM

The inductor RMS current in CCM is calculated as follows:

$$I_{Lc \, RMS\_c}(q) = \sqrt{\frac{2}{T_s} \left[\int_0^{\Delta t_3} \left(I_{Lc\_c} \frac{t}{\Delta t_3} \right)^2 dt + \int_0^{\Delta t_4} \left(I_{Lc\_c} + (I_1 - I_{Lc\_c}) \frac{t}{\Delta t_4} \right)^2 dt \right]}$$

$$(7.45)$$

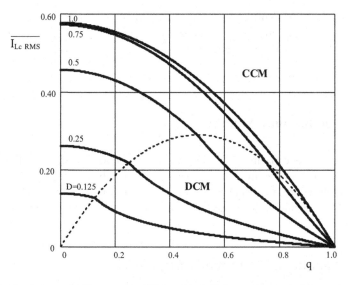

Fig. 7.23 The inductor RMS current in CCM and in DCM, as a function of the static gain (q), taking the duty cycle (D) as a parameter

Normalizing (7.45) gives:

$$
\overline{I_{Lc\,RMS\_C}(q)} = \frac{I_{Lc\,RMS\_C}4f_s\,L_c}{V_i} = \frac{(1-q)^2\,(D+q)^3 + (1+q)^2\,(D-q)^3}{6}
$$
$$
+\,\frac{4(1-D)(D-q^2)^2}{3}
\tag{7.46}
$$

(B) DCM

The inductor RMS current in DCM is calculated as follows:

$$
I_{Lc\,RMS\_D}(q) = \sqrt{\frac{2}{T_s}\left(\int_0^{\Delta T}\left(I_{Lc\_D}\frac{t}{\Delta T}\right)^2 dt + \int_0^{\Delta t_2}\left(I_{Lc\_D} - I_{Lc\_D}\frac{t}{\Delta t_2}\right)^2 dt\right)}
\tag{7.47}
$$

Normalizing (7.47) gives:

$$
\overline{I_{Lc\,RMS\_D}(q)} = \frac{I_{Lc\,RMS\_D}4f_s\,L_c}{V_i} = 2\sqrt{2}D\,(1-q)\sqrt{\frac{D}{6q}}
\tag{7.48}
$$

The inductor RMS current for CCM and DCM as well for the critical conduction mode (dashed line, making D = q) are presented in Fig. 7.23. As it can be observed, the smaller static gain (wider soft commutation range) the higher the inductor RMS current.

7.4 Commutation Analysis

In this section, the capacitors are added in parallel to the switches and dead time is considered, so the soft commutation phenomenon may be analyzed, adding another four intervals to one switching period in CCM and in DCM. If properly designed, the capacitors enable zero voltage switching (ZVS) in a wide load range.

7.4.1 Continuous Conduction Mode (CCM)

In CCM the leading leg (S_1 and S_3) commutates with the inductor peak current I_{Lc}_C and the lagging leg (S_2 and S_4,) commutates with a smaller current I_1 (Fig. 7.10). Another four time intervals of operation are added to the ones presented Sect. 7.2.1, these take place during the switches dead time. During the dead time intervals, that are much shorter than the other six, the inductor current is considered to remain constant and the capacitors in parallel to the switches must be fully charged/discharged in order to achieve soft commutation.

Figure 7.24 presents the first commutation time interval that takes place between time interval Δt_1 and Δt_2 of Sect. 7.2.1, when switch S_2 is turned off but switch S_4 is not yet gated on (dead time). The inductor current during this time interval is I_1 and half of this current ($I_1/2$) flows through each one of the capacitors, charging C_2 from

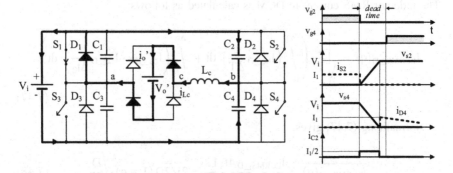

Fig. 7.24 Topological state and main waveforms for the first commutation time interval

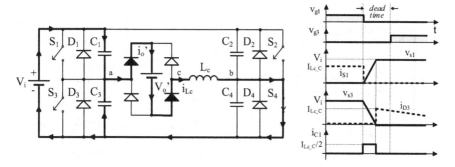

Fig. 7.25 Topological state and main waveforms for the second commutation time interval

zero to V_i and discharging C_4 from V_i to zero, linearly. Figure 7.24 also shows the main waveforms where the soft commutation of switches S_2 and S_4 may be observed. As soon as capacitor C_4 is discharged, diode D_4 conducts along with diode D_1 (time interval 2 of Sect. 7.2.1). To ensure soft commutation, the capacitors charge/discharge must finish before the dead time ends.

The second commutation time interval, shown in Fig. 7.25, takes place between time interval Δt_3 and Δt_4 of Sect. 7.2.1, when switch S_1 is turned off but switch S_3 is not yet gated on (dead time). The inductor current during this time interval is at its maximum value, I_{Lc}_C, and half of this current ($I_{Lc}\_C/2$) flows through each one of the capacitors, charging C_1 from zero to V_i and discharging C_3 from V_i to zero, linearly. As the inductor peak current is bigger than I_1, the charge/discharge of capacitors C_1 and C_3 are faster than the charge/discharge of capacitors C_2 and C_4. Figure 7.25 also shows the main waveforms where the soft commutation of switches S_1 and S_3 may be observed. At the instant capacitor C_3 is fully discharged,

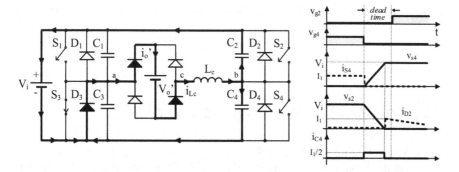

Fig. 7.26 Topological state and main waveforms for the third commutation time interval

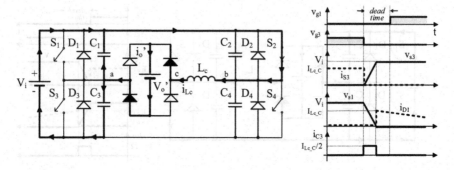

Fig. 7.27 Topological state and main waveforms for the fourth commutation time interval

diode D_3 starts conducting the inductor current along with switch S_4 (time interval Δt_4, Sect. 7.2.1).

Figure 7.26 shows the topological state and the main waveform for the third commutation time interval that takes place between time interval Δt_4 and Δt_5 of Sect. 7.2.1, when switch S_4 is turned off but switch S_2 is not yet gated on (dead time). This time interval is similar to the first one shown in Fig. 7.24.

The fourth commutation time interval, presented in Fig. 7.27, takes place between time interval Δt_6 and Δt_1 of Sect. 7.2.1, when switch S_3 is turned off but switch S_1 is not yet gated on (dead time). This time interval is similar to the second one shown in Fig. 7.25.

As it may be noticed in the description of the four commutation time intervals, the lagging leg commutates with a smaller current (I_1), so its commutation is more critical. It is not possible to ensure soft commutation in both legs in all the CCM load range, because I_1 reaches zero at the critical conduction mode, and soft commutation of switches S_2 and S_4 is no longer possible. For a given dead time and load range, the capacitors must be calculated for the critical leg, as follows:

$$C_1 = C_2 = C_3 = C_4 = \frac{I_1 \, t_d}{V_i} \tag{7.49}$$

7.4.2 Discontinuous Conduction Mode (DCM)

In DCM the leading leg (S_1 and S_3) commutates with the inductor peak current I_{Lc_D} and the lagging leg (S_2 and S_4,) has a dissipative commutation, because the inductor current is zero on its commutation, so no energy is available to charge/discharge the capacitors.

7.5 Simplified Design Methodology and an Example of the Commutation Parameters

In this section a design methodology and an example are presented according to the mathematical analysis presented in the previous sections. The converter is designed according to the specifications presented in Table 7.1.

A low static gain (q) of 0.4 is chosen, so a wider load range in CCM is obtained, resulting in a wider soft commutation load range. The output DC voltage referred to the primary side of the transformer (V'_o) is calculated as follows:

$$V'_o = V_i\, q = 400 \times 0.4 = 160\ V$$

The transformer turns ratio (n) and the output current referred to the primary side of the transformer (I'_o) are given by:

$$n = \frac{V'_o}{V_o} = \frac{160}{50} = 3.2$$

$$I'_o = \frac{I_o}{n} = \frac{10}{3.2} = 3.125\ A$$

Assuming that in the nominal power the duty cycle (D) is 0.9, the normalized output current referred to the primary side of the transformer is calculated:

$$\overline{I'_o{}_C} = \frac{I'_o\,C\,4f_s\,L_c}{V_i} = \frac{D(2-D)-q^2}{2} = \frac{0.9 \times (2-0.9) - 0.4^2}{2} = 0.415$$

Resulting in an inductance of:

$$L_c = \frac{0.415 \times 400}{3.125 \times 4 \times 40 \times 10^3} = 332\ \mu H$$

Table 7.1 Design Specifications

Input DC voltage (V_i)	400 V
Output DC voltage (V_o)	50 V
Output DC current (I_o)	10 A
Output power (P_o)	500 W
Switching frequency (f_s)	40 kHz
Dead time (t_d)	500 ns
Leading leg (S_1 and S_3) soft commutation range	30–100% P_o

The inductor RMS current is calculated as follows:

$$\overline{I_{Lc\,RMS\_C}} = \frac{(1-q)^2(D+q)^3 + (1+q)^2(D-q)^3}{6}$$
$$+ \frac{4(1-D)(D-q^2)^2}{3} = 0.496$$

$$I_{Lc\,RMS\_C} = \frac{\overline{I_{Lc\,RMS\_C}}V_i}{4f_s\,L_c} = \frac{400}{4 \times 40 \times 10^3 \times 332 \times 10^{-6}} = 3.732\ A$$

The switches S_1 e S_3 commutation current (I_{Lc}_C) as well as switches S_2 e S_4 commutation current (I_1), for nominal power, are given by:

$$\overline{I_{Lc\_C}} = (1-q) \times (D+q) = 0.78$$

$$I_{Lc\_C} = \frac{\overline{I_{Lc\_C}}V_i}{4f_s\,L_c} = \frac{0.78 \times 400}{4 \times 40 \times 10^3 \times 332 \times 10^{-6}} = 5.873\ A$$

$$\overline{I_1} = (1+q) \times (D-q) = 0.7$$

$$I_1 = \frac{\overline{I_1}\,V_i}{4f_s\,L_c} = \frac{0.7 \times 400}{4 \times 40 \times 10^3 \times 332 \times 10^{-6}} = 5.271\ A$$

The limit between CCM and DCM (D = q) is:

$$I'_{oL} = \frac{q(2-q) - q^2}{2} \times \frac{V_i}{4f_s\,L_c}$$

$$I'_{oL} = \frac{0.4 \times (2-0.4) - 0.4^2}{2} \times \frac{400}{4 \times 40 \times 10^3 \times 332 \times 10^{-6}} = 1.807\ A$$

Resulting in an output power of:

$$P_{oL} = I'_{oL} \times V'_o = 1.807 \times 160 = 289.16\ W$$

The capacitors in parallel to the switches has to be calculated to ensure soft commutation up to 30% of the nominal power (150 W) in the leading leg. As the critical conduction mode is on 289.16 W, the lagging leg (S_2 and S_4) will have a dissipative commutation on 30% of the nominal power, because the inductor current is already zero and there is no energy to charge/discharge the capacitors (C_2 and C_4). To ensure soft commutation at the leading leg (S_1 and S_3) on 30% of the nominal power, the capacitors have to be calculated as follows:

$$C_1 = C_2 = C_3 = C_4 = C = \frac{I_{Lc\_D}}{2} \times \frac{t_d}{V_i} = \frac{2.44}{2} \times \frac{500 \times 10^{-9}}{400} \cong 1.5\ nF$$

With these capacitors, the lagging leg soft commutation is achieved up to 70% of the nominal power, as calculated bellow:

$$I_{1\_L} = \frac{2CV_i}{t_d} = \frac{2 \times 1.5 \times 10^{-9} \times 400}{500 \times 10^{-9}} = 2.44 \text{ A}$$

$$\overline{I_{1\_L}} = \frac{I_{1\_L} 4L_c f_s}{V_1} = \frac{2.44 \times 4 \times 332 \times 10^{-6} \times 40 \times 10^3}{400} = 0.324$$

7.6 Simulation Results

The FB ZVS-PWM presented in Fig. 7.28 is simulated at the specified operation point to verify the analysis and the calculation presented in the design example of Sect. 7.5, with a dead time of 500 ns. Figure 7.29 presents the AC voltages v_{ab}, v_{ac} and v_{bc}, as well as the inductor current i_{Lc} and Table 7.2 presents the theoretical and simulated parameters, for nominal power.

A detail of the soft commutation in the nominal power for switches S_1 (leading leg) and S_2 (lagging leg) are presented in Figs. 7.30 and 7.31. As the capacitors were designed to achieve soft commutation in the leading leg at 30% of the nominal power, for both legs the capacitors charge/discharge finishes before the dead time ends. As the lagging leg has a smaller current ($I_1/2$) to charge/discharge the capacitors, the commutation takes longer than in the leading leg.

Figure 7.32 shows a detail of switch S_1 soft commutation at 30% of the nominal power. As it may be noticed the charge of capacitor C_1 takes all the dead time. Below 30% of the nominal power switch S_1 commutation is going to be dissipative.

Figure 7.33 the soft commutation limit for switch S_2 at 70% of the nominal power can be observed. The capacitor C_2 charge takes all the dead time. Below 70% of the nominal power the lagging leg has a dissipative commutation.

Fig. 7.28 Simulated converter

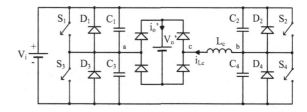

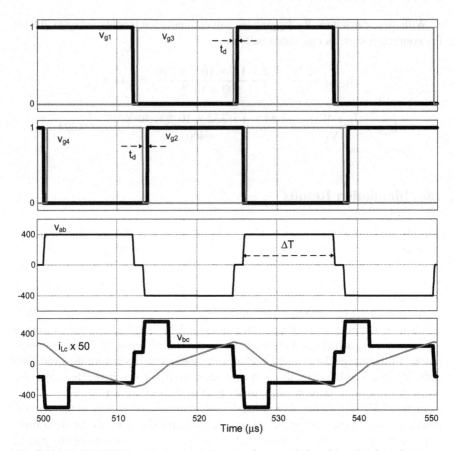

Fig. 7.29 FB ZVS-PWM converter simulation waveforms: switches drive signals, voltages v_{ab} and v_{bc} and inductor L_c current

Table 7.2 Theoretical and simulated results in CCM		Theoretical	Simulated
	I'_{oC} [A]	3.125	3.123
	I_1 [A]	5.271	5.286
	I_{Lc_C} [A]	5.873	5.849
	$I_{Lc\ RMS\_C}$ [A]	3.732	3.589
	Po [W]	500	499.783

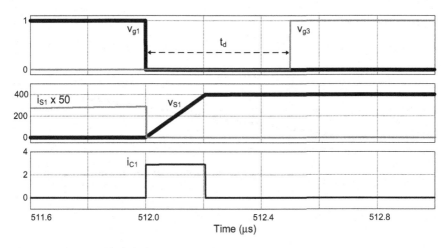

Fig. 7.30 Switch S_1 soft commutation at nominal power: S_1 and S_3 drive signal, voltage and current at switch S_1 and capacitor C_1 current at nominal power

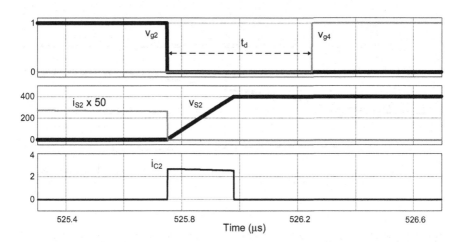

Fig. 7.31 Switch S_2 soft commutation at nominal power: S_2 and S_4 drive signal, voltage and current at switch S_2 and capacitor C_2 current at nominal power

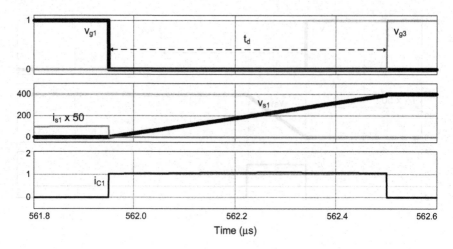

Fig. 7.32 Switch S_1 soft commutation at 30% of nominal power: S_1 and S_3 drive signal, voltage and current at switch S_1 and capacitor C_1 current

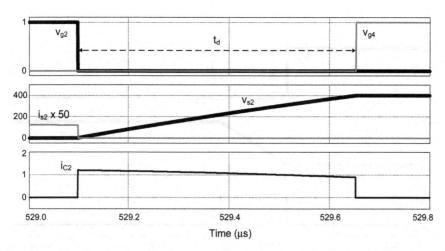

Fig. 7.33 Switch S_2 soft commutation at 70% of the nominal power: S_2 and S_4 drive signal, voltage and current at switch S_2 and capacitor C_2 current

7.7 Problems

(1) The FB-ZVS-PWM converter with capacitive output filter has the following specifications:

$V_i = 400$ V $V_o = 50$ V $\overline{I'_o} = 0.6$

$f_s = 50 \times 10^3$ Hz $I_o = 20$ A

(a) Calculate the transformer turns ratio.
(b) Calculate the inductance L_c.
(c) What is the duty cycle?
    Answers: (a) n = 4
    (b) $L_c = 120$ µH
    (c) D = 0.613

(2) The specification for a FB-ZVS-PWM converter with capacitive output filter are:

$V_i = 400$ V $V_o = 120$ V $N_s = N_p$ (n = 1)

(a) For the critical conduction mode calculate the duty cycle, the load average current and the output power.
(b) Draw the following curves: $I_o = f(D)$ e $P_o = f(D)$ para $0 \le D \le 1$.
    Answers: (a) D = 0.3, $I_o = 2.1$ A, $P_o = 252$ W
    $f_s = 100 \times 10^3$ Hz $L_c = 100$ µH

(3) A FB-ZVS-PWM converter with capacitive output filter have the following specifications:

$V_i = 400$ V, $V_o = 120$ V, $N_s = N_p$ (a = 1)
$f_s = 100 \times 10^3$ Hz $L_c = 100$ µH, D = 0.5

(a) Calculate the switches' commutation currents.
(b) Considering a dead time of 500 ns, calculate the commutation capacitance for each inverter leg.
    Answers: (a) $I_{Lc\_C} = 5.6$ A, $I_1 = 2.6$ A
    (b) $C_{13} = 3.5$ nF, $C_{24} = 1.625$ nF

(4) Describe the FB-PWM converter with capacitive output filter steps of operation, representing the topological states and the main waveforms, considering all components ideal, in CCM and in DCM.

(5) Obtain the expression of the normalized static gain as a function of the output current I_o, in CCM and in DCM.

(6) Describe the ZVS commutation steps of operation for the leading and lagging leg.

Fig. 7.34 Half bridge PWM converter

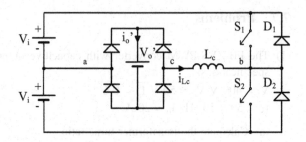

(7) The Half-Bridge PWM converter presented in Fig. 7.34 operates with switches S_1 and S_2 drive signals at 50% of duty cycle.

   (a) Considering that all components are ideal, describe the time intervals for the different topological states and present the main waveforms.

   (b) Obtain the average load current i'_o equation.

   (c) Obtain the inverter output characteristics $q = f\left(\overline{I'_o}\right)$, considering $q = V'_o/V_i$ and $I'_o = 4 L_K I_O/V_i$.

   (d) Considering the following specifications $V_i = 100$ V, $V_o = 50$ V, $f_s = 20$ kHz and $L_c = 100$ µH, calculate the average load current I_o and the output power P_o.

Reference

1. Barbi, I., Filho, W.A.: A non-resonant zero-voltage switching pulse-width modulated full-bridge DC-to-DC converter. In: IECON, pp. 1051–1056 (1990)

Chapter 8
Full-Bridge ZVS-PWM Converter with Inductive Output Filter

Nomenclature

V_i	Input DC voltage
V_o	Output DC voltage
P_o	Output power
C_o	Output filter capacitor
L_o	Output filter inductor
R_o	Output load resistor
ZVS	Zero voltage switching
ϕ	Angle between the leading leg and the lagging leg
q	Static gain
D	Duty cycle
D_{ef}	Effective duty cycle
f_s	Switching frequency
T_s	Switching period
t_d	Dead time
n	Transformer turns ratio
I'_o	Output DC current referred to the transformer primary side
$v'_o(V'_o)$	Output voltage referred to the transformer primary side and its average value
i_o	Output current
$I'_o\left(\overline{I'_o}\right)$	Average output current referred to the transformer primary and its normalized value
$I'_{o\,crit}$	Critical average output current referred to the primary side
S_1 and S_3	Switches in the leading leg
S_2 and S_4	Switches in the lagging leg
v_{g1}, v_{g2}, v_{g3} and v_{g4}	Switches S_1, S_2, S_3 and S_4 drive signals, respectively
D_1, D_2, D_3 and D_4	Diodes in anti-parallel to the switches (MOSFET—intrinsic diodes)
C_1, C_2, C_3 and C_4	Capacitors in parallel to the switches (MOSFET—intrinsic capacitors); $C = C_1 = C_2 = C_3 = C_4$
v_{C1}, v_{C2}, v_{C3} and v_{C4}	Capacitors voltage
v_C	Equivalent capacitors voltage

© Springer International Publishing AG, part of Springer Nature 2019
I. Barbi and F. Pöttker, *Soft Commutation Isolated DC-DC Converters*,
Power Systems, https://doi.org/10.1007/978-3-319-96178-1_8

L_c	Transformer leakage inductance or an additional inductor, if necessary
i_{Lc}	Inductor current
ω_o	Resonant frequency
z	Characteristic impedance
Δ and β	State plane angles
v_{ab}	Full bridge ac voltage, between points "a" and "b"
v_{cb}	Inductor voltage, between points "c" and "b"
v_{ac}	Voltage at the ac side of the rectifier, between points "a" and "c"
v_{S1}, v_{S2}, v_{S3} and v_{S4}	Voltage across switches
i_{S1}, i_{S2}, i_{S3} and i_{S4}	Current in the switches
i_{D1}, i_{D2}, i_{D3} and i_{D4}	Current in the diodes
i_{C1}, i_{C2}, i_{C3} and i_{C4}	Current in the capacitors
ΔT	Time interval which $v_{ab} = \pm V_i$
Δt_1	Time interval of the first step of operation in CCM
Δt_2	Time interval of the second step of operation in CCM
Δt_3	Time interval of the third step of operation in CCM
Δt_4	Time interval of the fourth step of operation in CCM
Δt_5	Time interval of the fifth step of operation in CCM
Δt_6	Time interval of the six step of operation in CCM
Δt_7	Time interval of the seventh step of operation in CCM
Δt_8	Time interval of the eighth step of operation in CCM
$I_{S13\,RMS}\left(\overline{I_{S13\,RMS}}\right)$	Switches S_1 and S_3 RMS current and its normalized value
$I_{S24\,RMS}\left(\overline{I_{S24\,RMS}}\right)$	Switches S_2 and S_4 RMS current in CCM and its normalized value
$I_{D13}\left(\overline{I_{D13}}\right)$	Diodes D_1 and D_3 average current and its normalized value
$I_{D24}\left(\overline{I_{D24}}\right)$	Diodes D_2 and D_4 average current and its normalized value

8.1 Introduction

In this chapter, the Full-Bridge Zero-Voltage-Switching Pulse-Width-Modulation Converter (FB-ZVS-PWM) with an Inductive Output Filter is studied. Such converter is similar to the FB-ZVS-PWM studied in Chap. 7. However, instead of a capacitive output filter, the output filter is inductive. This reduces the current ripple after the output rectifier, so it may be considered an ideal current source. As a consequence, in comparison to Chap. 7, the conduction losses are smaller, increasing significantly the performance, being suitable for even higher power applications. The chapter is organized to present at first the topology of the converter and its basic operation (without soft commutation) followed by a detailed analysis of the Zero Voltage Switching operation and a design example.

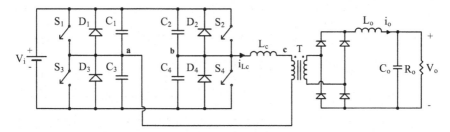

Fig. 8.1 Full-bridge ZVS-PWM with inductive output filter

The schematics of the FB-ZVS-PWM Converter with inductive output filter, shown in Fig. 8.1, consists of a single phase full bridge inverter, an inductor, an isolating transformer, a full bridge diode rectifier and an inductive output filter. Diodes (D_1, D_2, D_3 and D_4) in anti-parallel to the switches (S_1, S_2, S_3 and S_4) and the parallel capacitors (C_1, C_2, C_3 and C_4) can be the parasitic components of a MOSFET or additional components added in parallel to the switches, as in case of IGBTs. The inductance L_c may be the transformer leakage inductance or an additional one, if necessary. The main role of this inductor is to provide energy to charge and discharge the commutation capacitors at the instant of the commutation so that soft commutation can be achieved.

This converter operates with a fixed switching frequency and the voltage across the switches is limited to the DC link voltage (V_i). To achieve zero voltage switching (ZVS) the two legs operate with a phase-shift modulation, allowing the capacitors to be discharged and the antiparallel diodes to conduct prior to the switches conduction. At light load conditions, switches ZVS commutation may not be achieved, because of the insufficient energy stored in the inductor L_c.

8.2   Converter Operation

In this section, the converter represented by the equivalent circuit shown in Fig. 8.2 without the commutation capacitors is analyzed (the soft commutation analysis is presented in Sect. 8.3). To simplify, the following assumptions are made:

- all components are considered ideal;
- the converter is on steady-state operation;
- the output filter is represented as a DC current source I_o', whose value is the output current referred to the primary side of the transformer;
- current that flows through the switches is unidirectional, i.e. the switches allow the current to flow only in the direction of the arrow;
- the diodes in anti-parallel to the switches conduct separately from the switches, so the analysis is not limited to MOSFET's.

Fig. 8.2 Equivalent circuit of the full-bridge ZVS PWM converter

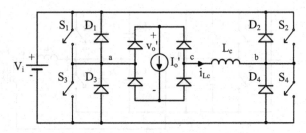

Fig. 8.3 Phase shift PWM used to control the FB ZVS-PWM

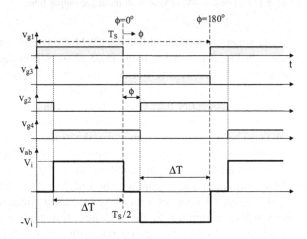

The basic principle of the phase shift PWM used to control this converter is illustrated in Fig. 8.3. The switches drive signals are presented for both the leading (S_1 and S_3) and the lagging leg (S_2 and S_4), along with the voltage v_{ab}. No dead time is considered at this point.

In the phase shift modulation, the switching frequency is fixed and the switches of each leg are gated complementary with duty cycle equal to 50%. The lagging leg drive signals are shifted of an angle ϕ in relation with to the leading leg drive signals, allowing the power transfer to be controlled. If $\phi = 0°$, v_{ab} has its maximum RMS value since switches S_1 and S_4 always conduct at the same time ($v_{ab} = V_i$), and switches S_2 and S_3 also conduct at the same time ($v_{ab} = -V_i$). When ϕ increases the v_{ab} RMS value decreases because switches S_1 and S_2 as well as S_3 and S_4 conduct at the same time resulting in $v_{ab} = 0$. The ϕ angle varies from $0°$ (maximum power) to $180°$ (zero power).

The operation of the converter is described over one switching period that is divided into eight time intervals, each one representing a different topological state.

(A) Time Interval Δt_1 ($t_0 \leq t \leq t_1$)

The topological state during the time interval Δt_1, shown in Fig. 8.4, starts at $t = t_0$. The output current source is short circuited by the rectifier diodes. The inductor L_c current flows through diode D_1 and switch S_2. The voltage v_{ab} is zero.

Fig. 8.4 Topological state
for time interval Δt_1

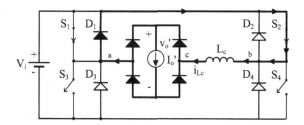

Fig. 8.5 Topological state
for time interval Δt_2

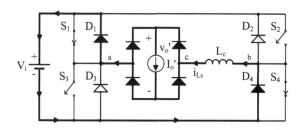

(B) Time Interval Δt_2 ($t_1 \leq t \leq t_2$)

This time interval, which topological state is shown in Fig. 8.5, starts at $t = t_1$, when S_4 is driven on and S_2 is driven off. As the inductor current is still negative $\left(i_{Lc}(t_1) = -I'_o\right)$, it flows through diodes D_1 and D_4 increasing linearly until it reaches zero. During this time interval, $v_{ab} = V_i$ and $v_{cb} = V_i$.

(C) Time Interval Δt_3 ($t_2 \leq t \leq t_3$)

Figure 8.6 presents the topological state for time interval Δt_3. It begins at $t = t_2$ when i_{Lc} reaches zero ($i_{Lc}(t_2) = 0$), turning off diodes D_1 and D_4 and enabling switches S_1 and S_4 to conduct, as the inductor current increases linearly. The AC voltage is $v_{ab} = V_i$ and the inductor voltage is $v_{cb} = V_i$.

(D) Time Interval Δt_4 ($t_3 \leq t \leq t_4$)

Figure 8.7 shows the topological state for time interval Δt_4. It begins at $t = t_3$ when i_{Lc} reaches I'_o $\left(i_{Lc}(t_3) = I'_o\right)$ so the output rectifier does not short circuit the output current anymore and energy is delivered to the load. The voltage $v_{ab} = V_i$.

Fig. 8.6 Topological state
for time interval Δt_3

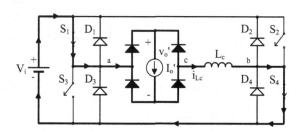

Fig. 8.7 Topological state
for time interval Δt_4

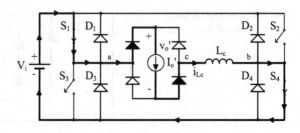

Fig. 8.8 Topological state
for time interval Δt_5

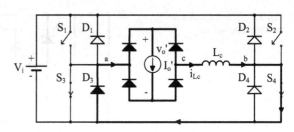

(E) Time Interval Δt_5 ($t_4 \leq t \leq t_5$)

The topological state for the time interval Δt_5 is shown in Fig. 8.8. It starts at $t = t_4$ when S_3 is driven on and S_1 is driven off. The output current source is short circuited by the rectifier diodes. The inductor L_c current flows through diode D_3 and switch S_4. The voltage v_{ab} is zero.

(F) Time Interval Δt_6 ($t_5 \leq t \leq t_6$)

This time interval, which topological state is shown in Fig. 8.9, starts at $t = t_5$, when S_2 is gated on and S_4 is gated off. As the inductor current is still positive ($i_{Lc}(t) = I'_o$), it flows through diodes D_2 and D_3 decreasing linearly until it reaches zero. The AC voltage is $v_{ab} = -V_i$ and the inductor voltage is $v_{cb} = -V_i$.

Fig. 8.9 Topological state
for time interval Δt_6

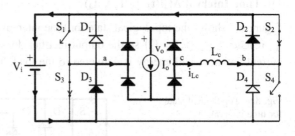

Fig. 8.10 Topological state
for time interval Δt_7

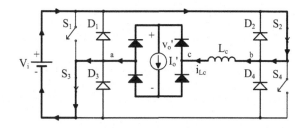

Fig. 8.11 Topological state
for time interval Δt_8

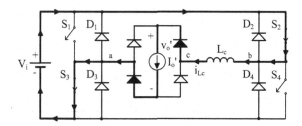

(G) Time Interval Δt_7 ($t_6 \le t \le t_7$)

Figure 8.10 presents the topological state for time interval Δt_7. It begins at $t = t_6$ when i_{Lc} reaches zero ($i_{Lc}(t) = 0$), turning off diodes D_2 and D_3, and enabling switches S_2 and S_3 to conduct, as the inductor current decreases linearly. The AC voltage is $v_{ab} = -V_i$ and the inductor voltage is $v_{cb} = -V_i$.

(H) Time Interval Δt_8 ($t_7 \le t \le t_8$)

Figure 8.11 presents the topological state for time interval Δt_8. It begins at $t = t_7$ when i_{Lc} reaches $-I_o'$. Thus, the output rectifier does not short circuit the output current anymore and energy is delivered to the load. The AC voltage is $v_{ab} = -V_i$.

The main waveforms and the time diagram are shown in Fig. 8.12.

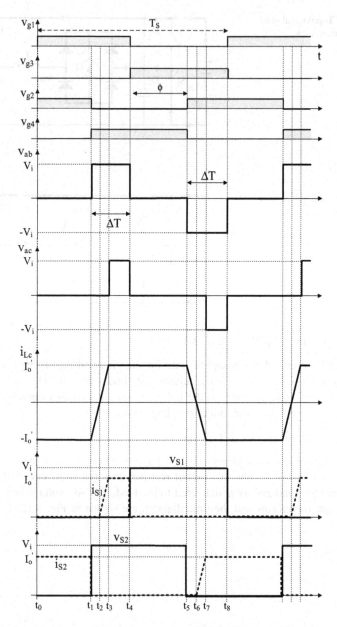

Fig. 8.12 Main Waveforms and time diagram of the FB ZVS-PWM converter over an operation cycle

8.3 Commutation Analysis

In this section, the capacitors are added in parallel to the switches and dead time is considered, so the soft commutation phenomenon may be analyzed, adding another four intervals to one switching period. If properly designed, the capacitors enable zero voltage switching (ZVS) in a wide load range.

The leading leg (S_1 and S_3) commutates when the output rectifier diodes are not short circuited, so the output current I_o' charge/discharge the capacitors in a linear way. The lagging leg (S_2 and S_4) commutates when the output rectifier is short circuited, and the charge/discharge of the commutation capacitors takes place in a resonant way, being more critical because only the inductor's (L_c) energy is available to charge/discharge the capacitors.

The four commutation time intervals take place during the switches dead-time, that are much shorter than the other eight time intervals. The capacitors in parallel to the switches must be fully charged/discharged in order to achieve soft commutation.

8.3.1 Leading Leg Commutation

During the leading leg commutations, the output rectifier is not short circuited, the output current I_o' charge/discharge the capacitors in a linear way.

Figure 8.13 shows the commutation time interval that takes place between time intervals Δt_4 and Δt_5 of Sect. 8.2, when switch S_1 is turned off but switch S_3 is not yet turned on (dead time). The inductor current during this time interval is the output current I_o' and half of this current flows through each of the capacitors, charging C_1 from zero to V_i and discharging C_3 from V_i to zero in a linear way. Figure 8.13 also shows the main waveforms where the soft commutation of switches S_1 and S_3 may be observed. As soon as the capacitor C_3 is discharged,

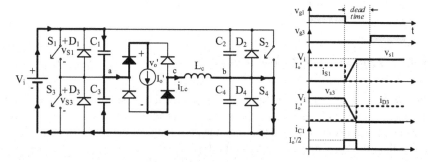

Fig. 8.13 Topological state and key waveforms during dead-time between gate signals of switches S_1 and S_3

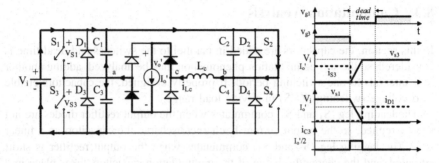

Fig. 8.14 Topological state and key waveforms during dead-time between gate signals of switches S_3 and S_1

diode D_3 conducts along with switch S_4 (time interval Δt_5 of Sect. 8.2). To ensure soft commutation, the capacitors charge/discharge must finish before the dead time ends.

Figure 8.14 shows the commutation time interval that takes place between time interval Δt_8 and Δt_1 of Sect. 8.2, when switch S_3 is driven off but switch S_1 is not yet driven on (dead time). The inductor current during this time interval is the output current $-I'_o$ and half of this current flows through each of the capacitors, charging C_3 from zero to V_i and discharging C_1 from V_i to zero in a linear way. Figure 8.14 also shows the main waveforms and the time diagram, where the soft commutation of switches S_1 and S_3 may be observed. As soon as the capacitor C_1 is discharged, diode D_1 conducts along with switch S_2 (time interval Δt_1 of Sect. 8.2). To ensure soft commutation, the capacitors charge/discharge must finish before the dead time ends.

To ensure soft commutation in the leading leg (non-critical commutation), for a given load current, the dead time t_d must satisfy the constraint given by Eq. (8.1).

$$t_d \geq \frac{C V_i}{I'_o} \tag{8.1}$$

where $C = C_1 = C_2 = C_3 = C_4$ is the commutation capacitor.

8.3.2   Lagging Leg Commutation

During the lagging leg commutation, the output rectifier is short circuited, and only the energy stored in inductance L_c is available to charge/discharge the commutation capacitors, in a resonant way. Therefore, the ZVS range of the switches of this leg is lower than the ZVS range of the leading leg.

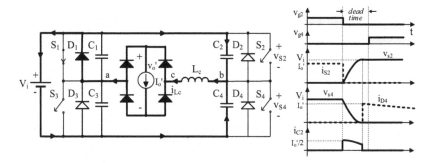

Fig. 8.15 Topological state and key waveforms during dead-time between gate signals of switches S_2 and S_4

Figure 8.15 shows topological state and relevant waveforms during dead-time between gate signals of switches S_2 and S_4 in which switch S_2 is gated off but switch S_4 is not gated on yet. As the output rectifiers diodes are short circuited at this time interval, only the energy stored in inductor L_c is available to the commutation process. To ensure soft commutation, the capacitors charge/discharge must finish before the dead time ends. It may be observed that the capacitors voltage and the inductor current evolve in a resonant way.

During the commutation process the inductor current flows through each one of the capacitors, charging one and discharging the other one. If the capacitor is discharged before the dead time ends, the soft commutation is achieved. Inductance L_c value is very important on this critical commutation, because only the energy stored on it is available to charge/discharge the capacitors. However, the choice of a high value of inductance, implies in a less effective duty cycle.

During the commutation time interval, the capacitor voltage and inductor current follow the equations bellow, considering $v_{C2}(0) = 0, v_{C4}(0) = V_i, i_{Lc}(0) = -I'_o$:

$$v_C(t) = z I'_o \, sen \, (\omega_o t) \tag{8.2}$$

$$i_{Lc}(t) = \frac{I'_o}{\omega_o^2} \cos \, (\omega_o t) \tag{8.3}$$

where $z = \sqrt{\frac{L_c}{2\,C}}$ and $\omega_o = \frac{1}{\sqrt{L_c\,2C}}$.

The state plane for the topological state of Fig. 8.15 is plotted in Fig. 8.16.

To ensure soft commutation, the constraint given by Eq. (8.4) must be satisfied:

$$z I'_o \geq V_i \tag{8.4}$$

Fig. 8.16 State plane for
topological state of Fig. 8.15

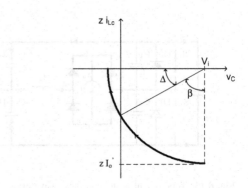

Thus,

$$I'_o \geq \sqrt{\frac{2C}{L_c}} V_i \tag{8.5}$$

According to Fig. 8.16 we have

$$\beta + \Delta = \frac{\pi}{2} \tag{8.6}$$

Hence,

$$\beta = \frac{\pi}{2} - \Delta = \omega \Delta t \tag{8.7}$$

After appropriate algebraic manipulation it can be found

$$\Delta t = \frac{\pi/2 - \Delta}{\omega} = \left(\frac{\pi}{2} - \Delta\right) \sqrt{2\,CL_c} \tag{8.8}$$

where Δt is the commutation time.
The angle Δ is given by

$$\Delta = \cos^{-1}\left(\frac{V_i}{z\,I'_o}\right) = \cos^{-1}\left(\frac{V_i}{I'_o} \sqrt{\frac{2C}{L_c}}\right) \tag{8.9}$$

Substituting (8.9) in (8.8) gives

$$\Delta t = \left[\frac{\pi}{2} - \cos^{-1}\left(\frac{V_i}{I'_o} \sqrt{\frac{2C}{L_c}}\right)\right] \sqrt{2\,CL_c} \tag{8.10}$$

The dead-time between the gate signals of the switches of the lagging leg must
be larger than the time interval Δt. Otherwise, the ZVS commutation of the power
switches is not achieved.

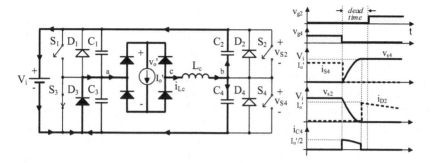

Fig. 8.17 Topological state and key waveforms during dead-time between gate signals of switches S_4 and S_2

Figure 8.17 shows the topological state and relevant waveforms of the other resonant commutation, that takes place between time intervals Δt_5 and Δt_6 of Sect. 8.2, during the dead-time between the gate signal of switches S_4 and S_2. In this commutation the capacitors voltages and the inductor current also evolve in a resonant way and only the energy stored in inductor L_c is available to the commutation process, since the output rectifier is short-circuited again. To ensure soft commutation, the capacitors charge/discharge must finish before the dead time ends.

8.4 Mathematical Analysis

8.4.1 Output Characteristics

The output characteristics of the FB-ZVS-PWM with an inductive output filter is obtained in this section. The commutation time intervals are disregarded because they are very short compared to the other time intervals and the inductor current may be considered constant during the commutation.

As the converter is symmetric in one switching period, the following time intervals are equal: $\Delta t_1 = \Delta t_5$, $\Delta t_2 = \Delta t_6$ and $\Delta t_3 = \Delta t_7$.

The time interval, which $v_{ab} = \pm V_i$, is defined as Δt, as shown in Fig. 8.12. Thus, Eqs. (8.11) and (8.12) may be written:

$$\Delta T = \Delta t_2 + \Delta t_3 + \Delta t_4 = \Delta t_6 + \Delta t_7 + \Delta t_8 \tag{8.11}$$

$$\frac{T_s}{2} = \Delta T + \Delta t_5 \tag{8.12}$$

The duty cycle is defined by

$$D = \frac{2\,\Delta T}{T_s} \qquad (8.13)$$

When the inductor L_c current evolves linearly in the time intervals Δt_2, Δt_3, Δt_6 and Δt_7, the output rectifier diodes are short circuited and the output voltage is zero, as it may be observed in Fig. 8.12. So, the power transferred to the load only takes place in the fourth and eight time intervals. The effective duty cycle is defined as follows:

$$\Delta t_4 = D_{ef}\,\frac{T_s}{2} \qquad (8.14)$$

As it can be noticed in the second and third time intervals, the equivalent circuit is the same and the voltage across inductor L_c is V_i, so these time intervals are equal and given by Eq. (8.15).

$$\Delta t_2 = \Delta t_3 = (D - D_{ef})\frac{T_s}{4} \qquad (8.15)$$

The first time interval be obtained as follows:

$$\Delta t_1 = (1 - D)\frac{T_s}{2} \qquad (8.16)$$

In the third time interval the current $i_{Lc}(t)$ is obtained from the following equation:

$$V_i = L_c\,\frac{di_{Lc}(t)}{dt} \qquad (8.17)$$

Applying the Laplace Transform yields

$$i_{Lc}(s) = \frac{V_i}{s^2\,L_c} \qquad (8.18)$$

Using the Inverse Laplace Transform the inductor current is obtained.

$$i_{Lc}(t) = \frac{V_i}{L_c}\,t \qquad (8.19)$$

This time interval ends when the inductor current reaches I_o', so time interval three may be calculated.

$$\Delta t_3 = \frac{I_o'\,L_c}{V_i} \qquad (8.20)$$

Substituting (8.14), (8.15) and (8.20) in (8.11), (8.21) is obtained.

Fig. 8.18 Output characteristic of the FB-ZVS-PWM with an inductive output filter

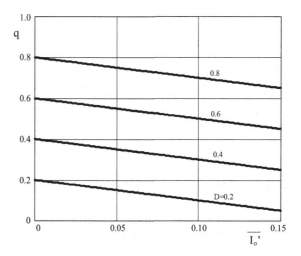

$$\Delta T = D \frac{T_s}{2} = \Delta t_3 + \Delta t_4 + \Delta t_5 = \frac{2 I'_o L_c}{V_i} + D_{ef} \frac{T_s}{2} \qquad (8.21)$$

Thus, the effective duty cycle is

$$D_{ef} = D - \frac{4 I'_o L_c f_s}{V_i} \qquad (8.22)$$

Defining

$$\overline{I'_o} = \frac{4 I'_o L_c f_s}{V_i} \qquad (8.23)$$

As it may be noticed in (8.22) and (8.23), $\overline{I'_o}$ represents the duty cycle loss due to the voltage drop in inductance L_c in time intervals Δt_2, Δt_3, Δt_6 and Δt_7, which is directly proportional to the output current $\overline{I'_o}$ and inductance L_c.

The average output voltage is defined by

$$V'_o = D_{ef} V_i \qquad (8.24)$$

Hence,

$$V'_o = \left(D - \frac{4 I'_o L_c f_s}{V_i} \right) V_i \qquad (8.25)$$

and

$$q = \frac{V'_o}{V_i} = D - \overline{I'_o} \tag{8.26}$$

where q is the converter static gain.

The FB-ZVS-PWM output characteristic, given by Eq. (8.26) is shown in Fig. 8.18.

8.4.2 RMS Switches Current

The switches S_1 and S_3 RMS current is calculating by integrating the inductor current in the third and fourth time interval (Sect. 8.2). Thus,

$$I_{S13\,RMS} = \sqrt{\frac{1}{T_s}\left[\int_0^{\Delta t_3}\left(I'_o\frac{t}{\Delta t_{32}}\right)^2 dt + \int_0^{\Delta t_4} I'^2_o\, dt\right]} = \frac{I'_o}{2}\sqrt{\frac{D+5\,D_{ef}}{3}} \tag{8.27}$$

Substitution of (8.22) into (8.27) and normalizing:

$$\overline{I_{S13\,RMS}} = \frac{I_{S13\,RMS}}{I'_o} = \frac{1}{2}\sqrt{2\,D - \frac{5}{3}\overline{I'_o}} \tag{8.28}$$

The switches S_2 and S_4 RMS current is calculated by integrating the inductor current in the seventh, eight and first time interval (Sect. 8.2). Hence,

$$I_{S24\,RMS} = \sqrt{\frac{1}{T_s}\left[\int_0^{\Delta t_7}\left(I'_o\frac{t}{\Delta t_{76}}\right)^2 dt + \int_0^{\Delta t_8} I'^2_o dt + \int_0^{\Delta t_1} I'^2_o\, dt\right]}$$

$$= I'_o\sqrt{\frac{-5\,(D - D_{ef}) + 6}{12}} \tag{8.29}$$

Substituting (8.22) in (8.29) and normalizing yields:

$$\overline{I_{S24\,RMS}} = \frac{I_{S24\,RMS}}{I'_o} = \frac{1}{2}\sqrt{2 - \frac{5}{3}\overline{I'_o}} \tag{8.30}$$

8.4.3 Average Diodes Current

The diodes D_1 and D_3 average current is calculated by integrating the inductor current in the first and third time intervals. Thus,

$$I_{D13} = \frac{1}{T_s} \left[\int_0^{\Delta t_1} I_o' \, dt + \int_0^{\Delta t_3} \left(I_o' - I_o' \frac{t}{\Delta t_3} \right) dt \right] = \frac{I_o'}{2} \left[1 - \frac{(D + D_{ef})}{2} \right] \quad (8.31)$$

After integration and appropriate algebraic manipulation, we find

$$\overline{I_{D13}} = \frac{I_{D13}}{I_o'} = \frac{1}{2} \left[1 - D + \frac{\overline{I_o'}}{2} \right] \quad (8.32)$$

Similarly, the average value of the diodes D_2 and D_4 current is determined by integration of the inductor current in the second time interval, according to Eq. (8.32).

$$I_{D24} = \frac{1}{T_s} \int_0^{\Delta t_2} \left(I_o' - \frac{\overline{I_o'}}{\Delta t_2} t \right) dt = I_o' \frac{(D - D_{ef})}{8} \quad (8.33)$$

Integration and algebraic manipulation gives

$$\overline{I_{D24}} = \frac{I_{D24}}{I_o'} = \frac{\overline{I_o'}}{8} \quad (8.34)$$

8.5 Simplified Design Example and Methodology

In this section, a simplified design example is presented, using the analysis results of previous sections. The converter specifications are given in Table 8.1.

The static gain is $q = 0.4$. The output DC voltage referred to the primary side of the transformer $\left(V_o' \right)$ is calculated as follows:

$$V_o' = V_i \, q = 400 \times 0.4 = 160 \text{ V}$$

Table 8.1 Design specifications

Input DC voltage (V_i)	400 V
Output DC voltage (V_o)	50 V
Output DC current (I_o)	10 A
Output power (P_o)	500 W
Switching frequency (f_s)	40 kHz

The transformer turns ratio (n) and the output current referred to the primary side of the transformer (I_o') are:

$$n = \frac{V_o'}{V_o} = \frac{160}{50} = 3.2$$

and

$$I_o' = \frac{I_o}{n} = \frac{10}{3.2} = 3.125 \text{ A}$$

Assuming that at nominal power the duty cycle reduction is 15%, the inductance L_c is:

$$L_c = \frac{\overline{I_o'} V_i}{4 f_s I_o'} = \frac{0.15 \times 400}{4 \times 40 \times 10^3 \times 3.125} = 120 \text{ μH}$$

The nominal and effective duty cycles are calculated as follows:

$$D_{nom} = \frac{V_o'}{V_i} + \overline{I_o'} = \frac{160}{400} + 0.15 = 0.55$$

$$D_{ef} = D_{nom} - \overline{I_o'} = 0.55 - 0.15 = 0.4$$

The switches RMS current and the antiparallel diodes average current, for nominal power, are:

$$\overline{I_{S13\,RMS}} = \frac{I_{S13\,RMS}}{I_o'} = \frac{1}{2}\sqrt{2D - \frac{5}{3}\overline{I_o'}} = \frac{1}{2}\sqrt{2 \times 0.55 - \frac{5}{3}0.15} = 0.46$$

$$I_{S13\,RMS} = 0.46 \times 3.125 = 1.44 \text{ A}$$

$$\overline{I_{S24\,RMS}} = \frac{I_{S24\,RMS}}{I_o'} = \frac{1}{2}\sqrt{2 - \frac{5}{3}\overline{I_o'}} = \frac{1}{2}\sqrt{2 - \frac{5}{3}0.15} = 0.66$$

$$I_{S24\,RMS} = 0.66 \times 3.125 = 2.06 \text{ A}$$

$$\overline{I_{D13}} = \frac{I_{D13}}{I_o'} = \frac{1}{2}\left(1 - D + \frac{\overline{I_o'}}{2}\right) = \frac{1}{2}\left(1 - 0.55 + \frac{0.15}{2}\right) = 0.26$$

$$I_{D13} = 0.26 \times 3.125 = 0.82 \text{ A}$$

$$\overline{I_{D24}} = \frac{I_{D24}}{I'_o} = \frac{\overline{I'_o}}{8} = \frac{0.15}{8} = 0.02$$

$$I_{D24} = 0.02 \times 3.125 = 0.06 \text{ A}$$

Assuming that the capacitances of the commutation capacitors are $C = C_1 = C_2 = C_3 = C_4 = 0.5$ nF.

The lagging leg soft commutation limit implies that angle Δ is zero. So, the commutation time is:

$$\Delta t = \frac{\pi}{2} \sqrt{2\,CL_c} = \frac{\pi}{2} \times \sqrt{2 \times 0.5 \times 10^{-9} \times 120 \times 10^{-6}} = 544 \text{ ns}$$

As it has already been demonstrated in previous sections, the accomplishment of the ZVS commutation requires a dead-time larger than 544 ns.

As it may be noticed in Fig. 8.16, the soft commutation limit also implies that $zI_o = V_i$. So, the minimum output current that ensures soft commutation is

$$I'_{o\,crit} = V_i \sqrt{\frac{2C}{L_c}} = 400 \sqrt{\frac{2 \times 500 \times 10^{-12}}{120 \times 10^{-6}}} = 1.15 \text{ A}$$

This value is 37% of the rated current, which implies in ZVS commutation for power range between 37% and 100% of the rated power.

8.6 Simulation Results

The FB ZVS-PWM with an inductive output filter presented in Fig. 8.19 is simulated to validate the analysis, for a dead time of 550 ns. Figure 8.20 presents the voltages v_{ab}, v'_o, as well as the inductor voltage and current. Table 8.2 presents theoretical and simulated parameters and component stresses, validating the mathematical analysis.

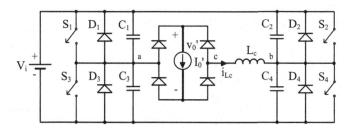

Fig. 8.19 Simulated converter

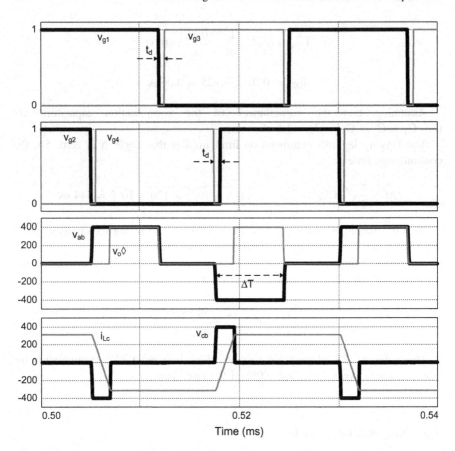

Fig. 8.20 FB-ZVS-PWM converter simulation waveforms for nominal power: switches drive signals, voltages v_{ab}, v_o' and v_{cb}, current i_{Lc}

Figures 8.21 and 8.22 present the leading leg and the lagging leg commutation at nominal power. As it may be noticed soft commutation is achieved.

Table 8.2 Theoretical and simulated results

	Theoretical	Simulated
V_o' [V]	160	160.08
$I_{S13\ RMS\_}$ [A]	1.44	1.44
$I_{S24\ RMS}$ [A]	2.06	2.07
I_{D13} [A]	0.82	0.76
I_{D24} [A]	0.06	0.058
P_o [W]	500	500.25

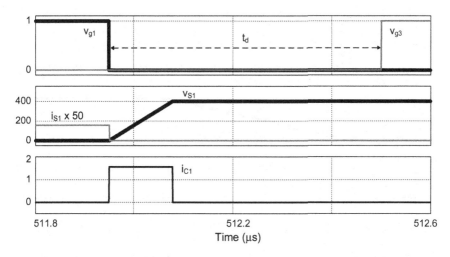

Fig. 8.21 Switch S_1 soft commutation at nominal power: S_1 and S_3 drive signals, voltage and current at switch S_1 and capacitor C_1 current

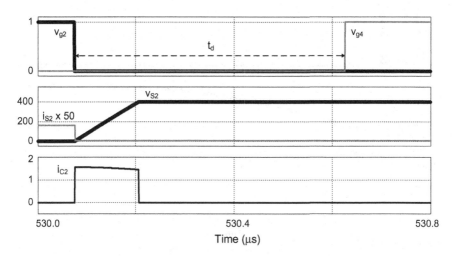

Fig. 8.22 Switch S_2 soft commutation at nominal power: S_2 and S_4 drive signals, voltage and current at switch S_2 and capacitor C_2 current

Figures 8.23 and 8.24 present the leading leg and the lagging leg commutation at the minimum power. As it may be noticed soft commutation is on its limit at the lagging leg.

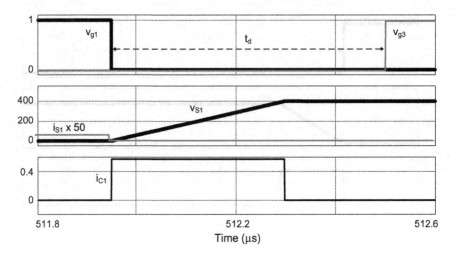

Fig. 8.23 Switch S_1 soft commutation at minimum power: S_1 and S_3 drive signals, voltage and current at switch S_1 and capacitor C_1 current

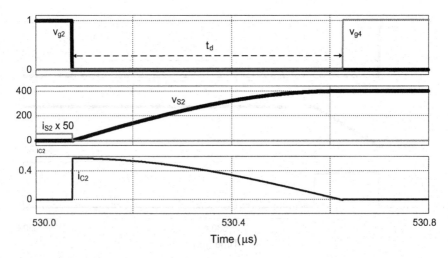

Fig. 8.24 Switch S_2 soft commutation at the minimum power: S_2 and S_4 drive signals, voltage and current at switch S_2 and capacitor C_2 current

8.7 Problems

(1) The FB-PWM converter shown in Fig. 8.25 has a capacitor C to block v_{ab} average value caused by not ideal components. Considering the following specifications, draw the capacitor C voltage (v_C) and calculate its peak value, considering all components ideal.

$$V_i = 400 \text{ V} \quad I_o' = 10 \text{ A} \quad f_s = 20 \text{ kHz} \quad L_c = 40 \text{ μH} \quad C = 10 \text{ μF}$$

Answer: $v_{C \text{ peak}} = 12$ V

(2) For the ideal FB-PWM converter shown in Fig. 8.26: (a) obtain the expression of the output voltage V_o and (b) calculate the average output voltage. The parameters of the converter are:

$$V_i = 400 \text{ V} \quad L_o = 2 \text{ mH} \quad C_o = 100 \text{ μF} \quad N_s = N_p (n = 1)$$
$$R_o = 20 \text{ Ω} \quad f_s = 40 \text{ kHz} \quad L_c = 40 \text{ μH} \quad D = 0.75 (f = 45°)$$

Answers: (a) $V_o = (DR_o) / (R_o + 4 L_c f_s)$; (b) $V_o = 227.27$ V

(3) A FB-PWM converter and its drive signals are shown in Fig. 8.27. Due to the asymmetric gate drive signals, the voltage v_{ab} contains a DC component, which is blocked by the capacitor C. Calculate this DC voltage component, considering the following parameters:

$$I_o = 10 \text{ A} \quad C = 10 \text{ μF} \quad \Delta t = 1 \text{ μs} \quad N_s = N_p (a = 1)$$
$$f_s = 40 \text{ kHz} \quad L_c = 25 \text{ μH} \quad L_m = 1 \text{ mH}$$

Answer: $V_c = 8$ V

(4) For the FB-ZVS-PWM converter shown in Fig. 8.28 with the specifications given below, calculate: (a) the minimum value of I_o' and (b) the dead-time required to enable ZVS commutation on all switches.

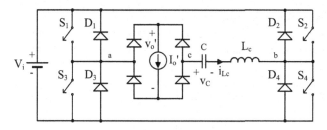

Fig. 8.25 Full bridge ZVS PWM converter

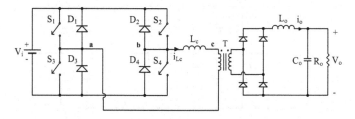

Fig. 8.26 Full bridge PWM

Fig. 8.27 Full bridge
ZVS-PWM converter and its
drive signals

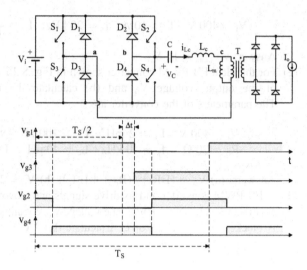

$$V_i = 400 \text{ V} \quad C = C_1 = C_2 = C_3 = C_4 = 0.5 \text{ nF} \quad f_s = 40 \text{ kHz}$$
$$L_c = 10 \text{ μH} \quad \Delta_t = 300 \text{ ns}$$

Answers: (a) $I'_0 = 2A$; (b) $t_d = 314.2$ ns.

(5) The FB-PWM converter is shown in Fig. 8.29 with the transformer magnetizing inductance.

  (a) Describe the converter operation during one switching period as well as the
      main waveforms;
  (b) Obtain the static gain expression;
  (c) Obtain the magnetizing inductance peak current expression;
  (d) Obtain the expression to calculate the switches current at the commutation
      instant.

Answers: (b) $\quad q = \dfrac{V_o}{V_i} = \left(\dfrac{L_m}{L_c + L_m}\right) \times \left(D - \dfrac{4f_s L_c I_o}{V_i}\right)$;

(c) $I_{m\,peak} = \dfrac{V_i}{4f_s(L_c + L_m)} \times \left(D - \dfrac{4f_s L_c I_o}{V_i}\right)$; (d) $I_{s\,peak} = I_o + I_{m\,peak}$

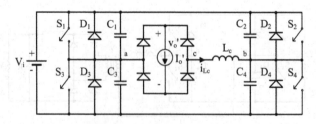

Fig. 8.28 Full bridge ZVS-PWM converter

Fig. 8.29 Full bridge ZVS-PWM converter with transformer magnetizing inductance

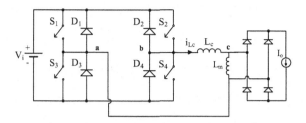

(6) The FB-ZVS-PWM converter shown in Fig. 8.29 has the following parameters:

$$V_i = 400 \text{ V} \quad I_o = 10 \text{ A} \quad f_s = 20 \text{ kHz}$$
$$L_c = 50 \text{ μH} \quad L_m = 500 \text{ μH} \quad D = 0.8$$

Find (a) the average value of the output voltage V_o, (b) the peak value of the magnetizing current and (c) the peak value of the current in a power switch at the instant of the commutation.

Answers: (a) $V_o = 254.54$ V; (b) $I_{m \text{ peak}} = 6.36$ A; (c) $I_{s \text{ peak}} = 16.36$ A

Chapter 9
Three-Level Neutral Point Clamped ZVS-PWM Converter

Nomenclature

V_i	Input DC voltage
V_o	Output DC voltage
P_o	Output power
C_o	Output filter capacitor
L_o	Output filter inductor
R_o	Output load resistor
PWM	Pulse width modulation
ZVS	Zero voltage switching
q	Static gain
D	Duty cycle
D_{nom}	Nominal duty cycle
D_{ef}	Effective duty cycle
ΔD	Duty cycle loss
f_s	Switching frequency
T_s	Switching period
t_d	Dead time
T	Transformer
n	Transformer turns ratio
v_o' (V_o')	Output voltage referred to the transformer primary side and its average value
i_o	Output current
I_o' $\left(\overline{I_o'}\right)$	Average output current referred to the primary side and its normalized value
$I_{o\,crit}'$	Critical average output current referred to the primary side
S_1, S_2, S_3 and S_4	Switches

I. Barbi and F. Pöttker, *Soft Commutation Isolated DC-DC Converters*,
Power Systems, https://doi.org/10.1007/978-3-319-96178-1_9

v_{g1}, v_{g2}, v_{g3} and v_{g4}	Switches S_1, S_2, S_3 and S_4 drive signals, respectively
D_1, D_2, D_3 and D_4	Diodes in anti-parallel to the switches (MOSFET—intrinsic diodes)
D_{C1}, D_{C2}	Clamping diodes
$C = C_1 = C_2 = C_3 = C_4$	Capacitors in parallel to the switches (MOSFET—intrinsic capacitors)
L_r	Resonant inductor
C_r	Resonant capacitor
$i_{Lr\ peak}$	Resonant inductor peak current
$v_{Cr\ peak}$	Resonant inductor peak voltage
L_c	Commutation inductor (may be the transformer leakage inductance or an additional inductor, if necessary)
i_{Lc}	Commutation inductor current
ω_o	Resonant frequency
v_{C1}, v_{C2}, v_{C3} and v_{C4}	Capacitors voltage
v_{ab}	Ac voltage, between points "a" and "b"
v_{S1}, v_{S2}, v_{S3} and v_{S4}	Voltage across the switches
i_{S1}, i_{S2}, i_{S3} and i_{S4}	Current in the switches
i_{C1}, i_{C2}, i_{C3} and i_{C4}	Current in the capacitors
ΔT	Time interval in which $v_{ab} = \pm V_i$
Δt_2	Time interval of the first and fourth step of operation in DCM
Δt_{10}	Time interval of the first step of operation in CCM (t_1–t_0)
Δt_{21}	Time interval of the second step of operation in CCM (t_2–t_1)
Δt_{32}	Time interval of the third step of operation in CCM (t_3–t_2)
Δt_{43}	Time interval of the fourth step of operation in CCM (t_4–t_3)
Δt_{54}	Time interval of the fifth step of operation in CCM (t_5–t_4)
Δt_{65}	Time interval of the sixth step of operation in CCM (t_6–t_5)
Δt_{76}	Time interval of the seventh step of operation in CCM (t_7–t_6)
Δt_{87}	Time interval of the eighth step of operation in CCM (t_8–t_7)

$I_{S14\,RMS}\ \left(\overline{I_{S14\,RMS}}\right)$ Switches S_1 and S_4 RMS current and its normalized value

$I_{S23\,RMS}\ \left(\overline{I_{S23\,RMS}}\right)$ Switches S_2 and S_3 RMS current and its normalized value

$I_{D1234}\ \left(\overline{I_{D1234}}\right)$ Diodes D_1, D_2, D_3 and D_4 average current and its normalized value

$I_{DC12}\ \left(\overline{I_{DC12}}\right)$ Clamping diodes average current and its normalized value

9.1 Introduction

The converters studied in the previous chapters are not suitable for applications in which the DC bus voltage exceeds the maximum voltage that the power switches can withstand.

In order to overcome this drawback, in 1993 [1] the Three-Level Neutral Point Clamped ZVS-PWM converter, shown in Fig. 9.1, was proposed, which operates in a similar way to the FB ZVS-PWM converter with an inductive output filter presented in Chap. 8.

The commutation of the power switches and the output characteristics of the TL-ZVS-PWM converter are identical to the Full-Bridge ZVS-PWM converter. However, the power switches are submitted to a voltage equal to half of the DC bus voltage ($V_i/2$).

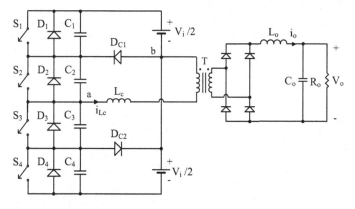

Fig. 9.1 Three-Level NPC ZVS-PWM converter with an inductive output filter

9.2   Circuit Operation

In this section, the ideal converter shown in Fig. 9.2, with all parameters referred to the transformer primary winding is analyzed. To simplify, the following assumptions are made:

- all components are considered ideal;
- the converter operates on steady-state condition;
- the output filter is represented as a DC current source I'_o, whose value is the output average current referred to the primary winding of the transformer;
- the converter is controlled by pulse width modulation (PWM), with switches S_2 and S_3 operating complementary with 50% duty cycle and switches S_1 and S_4 operating with variable duty cycle D to control the power transferred to the load.

The operation of the converter will be described for one switching period that is divided into eight time intervals, each one representing a different state of the switches.

(A)   Time Interval Δt_1 ($t_0 \leq t \leq t_1$)

The time interval Δt_1, shown in Fig. 9.3, starts at $t = t_0$ when the inductor current is equal I'_o. Switches S_1 and S_2 conduct the output current, the voltage v_{ab} is equal to $V_i/2$ and energy is being transferred to the load.

(B)   Time Interval Δt_2 ($t_1 \leq t \leq t_2$)

At the instant $t = t_1$ switch S_1 is turned off and the clamping diode D_{c1} starts to conduct the output current along with switch S_2, as shown in Fig. 9.4. During this time interval the voltage v_{ab} is zero and the output current source is short circuited by the rectifier diodes.

(C)   Time Interval Δt_3 ($t_2 \leq t \leq t_3$)

The time interval Δt_3 begins at $t = t_2$ when switch S_2 is turned off and switches S_3 and S_4 are gated on. As the inductor current is positive, the current flows through diodes D_3 and D_4, as it can be seen in Fig. 9.5. The output current source remains short circuited by the rectifier diodes. The voltage v_{ab} is equal to $-V_i/2$ and the L_c inductor current decreases linearly.

Fig. 9.2 Equivalent circuit of the ideal Three-Level NPC PWM converter

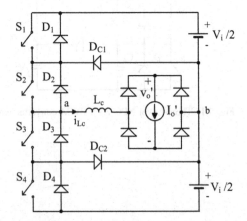

Fig. 9.3 Topological state during time interval Δt_1

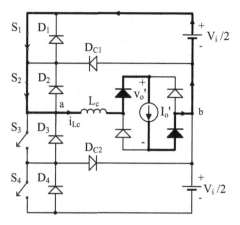

Fig. 9.4 Topological state during time interval Δt_2

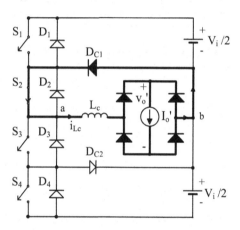

Fig. 9.5 Topological state during the time interval Δt_3

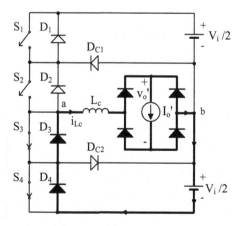

(D) Time Interval Δt_4 ($t_3 \leq t \leq t_4$)

This time interval begins at the instant t_3, when inductor current i_{Lc} reaches zero. Switches S_3 and S_4 start to conduct the inductor current i_{Lc}, that decreases linearly. The output current source remains short circuited by the rectifier diodes. The topological state for this time interval is shown in Fig. 9.6.

(E) Time Interval Δt_5 ($t_4 \leq t \leq t_5$)

The time interval Δt_5, shown in Fig. 9.7, starts at $t = t_4$, when the inductor current i_{Lc} reaches $-I_o'$. Switches S_1 and S_2 conduct the output current $-I_o'$, the voltage v_{ab} is equal to $-V_i/2$ and energy is transferred to the load.

(F) Time Interval Δt_6 ($t_5 \leq t \leq t_6$)

At $t = t_5$ the switch S_4 is turned off, the clamping diode D_{c2} starts to conduct the output current along with switch S_3, as shown in Fig. 9.8. As the voltage v_{ab} is zero, the output current source is short circuited again by the rectifier diodes.

Fig. 9.6 Topological state during the time interval Δt_4

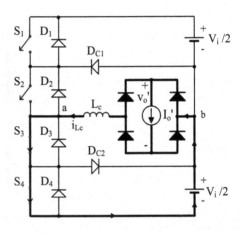

Fig. 9.7 Topological state during time interval Δt_5

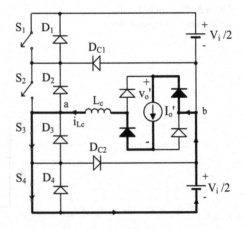

(G)   Time Interval Δt_7 ($t_6 \leq t \leq t_7$)

The time interval Δt_7 begins at $t = t_6$ when switch S_3 is turned off and switches S_1 and S_2 are gated on. As the inductor current i_{Lc} is negative, it flows through diodes D_1 and D_2, as shown in Fig. 9.9. The output current source is still short circuited by

Fig. 9.8 Topological state during time interval Δt_6

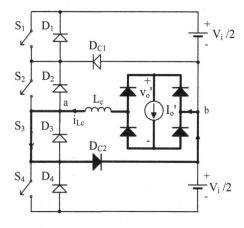

Fig. 9.9 Topological state during time interval Δt_7

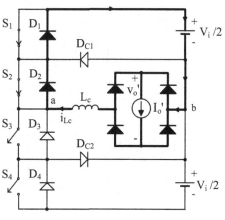

Fig. 9.10 Topological state during time interval Δt_8

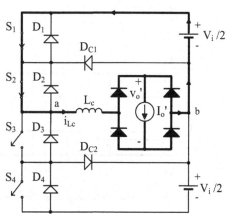

the rectifier diodes. The voltage v_{ab} is equal to $V_i/2$, increasing linearly the inductor current i_{Lc}.

(H) Time Interval Δt_8 ($t_7 \leq t \leq t_8$)

This time interval, shown in Fig. 9.10, begins at the instant t_7 when switches S_1 and S_2 start to conduct the inductor current i_{Lc}, that increases linearly. The output current source remains short circuited by the rectifier diodes.

The relevant waveforms and time diagram are shown in Fig. 9.11.

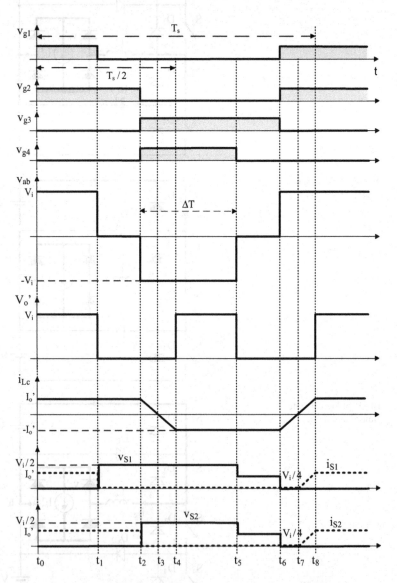

Fig. 9.11 Relevant waveforms and time diagram over one cycle of operation

9.3   Mathematical Analysis

(A) Output Characteristics

As the converter is symmetric in one switching period, we have: $\Delta t_{10} = \Delta t_{54}$, $\Delta t_{21} = \Delta t_{65}$, $\Delta t_{32} = \Delta t_{76}$, and $\Delta t_{43} = \Delta t_{87}$.

According to the time diagram shown in Fig. 9.11, the time intervals in which $v_{ab} = \pm V_i$ is given by

$$\Delta T = \Delta t_{32} + \Delta t_{43} + \Delta t_{54} = \Delta t_{10} + \Delta t_{76} + \Delta t_{87} \tag{9.1}$$

The half cycle of a switching period is given by

$$\frac{T_s}{2} = \Delta T + \Delta t_{21} \tag{9.2}$$

According to the duty-cycle definition, we have

$$D = \frac{\Delta T}{T_s/2} \tag{9.3}$$

During the time interval in which the inductor i_{Lc} current evolves linearly, the output voltage v_o' is zero So, no power is transferred to the load in this time interval. The effective duty cycle (D_{ef}) is defined by Eq. (9.4).

$$D_{ef} = \frac{\Delta t_{10}}{T_s/2} = \frac{\Delta t_{54}}{T_s/2} \tag{9.4}$$

The time intervals Δt_{32} and Δt_{21} are calculated as follows:

$$\Delta t_{32} = (D - D_{ef})\frac{T_s}{4} \tag{9.5}$$

$$\Delta t_{21} = (1 - D)\frac{T_s}{2} \tag{9.6}$$

Based on the topological state for time interval Δt_4 and Δt_8, we have

$$\frac{V_i}{2} = L_c \frac{di_{Lc}(t)}{dt} \tag{9.7}$$

Applying the Laplace transformation to Eq. (9.7) we find

$$i_{Lc}(s) = \frac{V_i/2}{s^2 L_c} \tag{9.8}$$

Applying the inverse Laplace transformation to Eq. (9.8) gives

$$i_{Lc}(t) = \frac{V_i/2}{L_c} t \qquad (9.9)$$

Time intervals Δt_{43} and Δt_{87} ends when the inductor current reaches I_o' and is given by Eqs. (9.10).

$$\Delta t_{43} = \Delta t_{87} = \frac{I_o' L_c}{V_i/2} \qquad (9.10)$$

Substituting (9.4) and (9.10) into (9.1) yields

$$\Delta T = D\frac{T_s}{2} = \Delta t_{32} + \Delta t_{43} + \Delta t_{54} = \frac{2 I_o' L_c}{V_i/2} + D_{ef}\frac{T_s}{2} \qquad (9.11)$$

Thus, from Eq. (9.11),

$$D_{ef} = D - \frac{4 I_o' L_c f_s}{V_i/2} \qquad (9.12)$$

Equation (9.12) can also be written in the compact form

$$D_{ef} = D - \overline{I_o'} \qquad (9.13)$$

where $\overline{I_o'} = \frac{4 I_o' L_c f_s}{V_i/2}$ is the duty-cycle loss caused by the series inductance L_c.

From the above equations we find Eq. (9.14) that represents the converter output voltage referred to the transformer primary winding.

$$V_o' = D_{ef}\frac{V_i}{2} = \left(D - \overline{I_o'}\right) \times \frac{V_i}{2} \qquad (9.14)$$

Consequently, the static gain (q) is given by:

$$q = \frac{V_o'}{V_i/2} = D - \overline{I_o'} \qquad (9.15)$$

The three-level ZVS-PWM converter theoretical output characteristics, given by Eq. (9.15) (static gain as a function of the normalized output current, taking the duty cycle as parameter) are shown in Fig. 9.12. As it may be noticed, the average output voltage (V_o') depends on the load current (I_o'), due to the voltage drop across inductor L_c.

Fig. 9.12 Three-level
ZVS-PWM converter output
characteristics

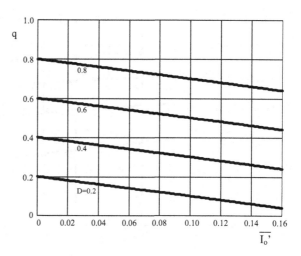

(B) Switches RMS Current

The effective or RMS current that circulates through switches S_1 and S_4 is given by

$$I_{S14\,RMS} = \sqrt{\frac{1}{T_s}\left[\int_0^{\Delta t_{87}}\left(I_o'\frac{t}{\Delta t_{87}}\right)^2 dt + \int_0^{\Delta t_{10}} I_o'^2\, dt\right]} \qquad (9.16)$$

After integration in relation to t, yields

$$I_{S14\,RMS} = \frac{I_o'}{2}\sqrt{\frac{D + 5\,D_{ef}}{3}} \qquad (9.17)$$

Substitution of Eqs. (9.12) into (9.17), with appropriate algebraic manipulation gives the normalized RMS current in switches S_1 and S_4:

$$\overline{I_{S14\,RMS}} = \frac{I_{S14\,RMS}}{I_o'} = \frac{1}{2}\sqrt{2\,D - \frac{5\,I_o'}{3}} \qquad (9.18)$$

Similarly, the RMS value of current in switches S_2 and S_3 is given by

$$I_{S23\,RMS} = \sqrt{\frac{1}{T_s}\left(\int_0^{\Delta t_{87}}\left(I_o'\frac{t}{\Delta t_{87}}\right)^2 dt + \int_0^{\Delta t_{10}} I_o'^2\, dt + \int_0^{\Delta t_{21}} I_o'^2\, dt\right)} \qquad (9.19)$$

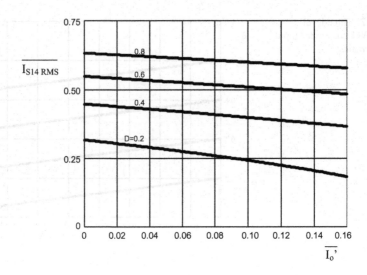

Fig. 9.13 Switches S_1 and S_4 RMS current, as a function of the normalized output current, taking D as a parameter

After integration in relation to time and algebraic manipulation it can be found

$$I_{S23\,RMS} = I'_o \sqrt{\frac{6 - 5\,(D - D_{ef})}{12}} \tag{9.20}$$

Substitution of Eq. (9.12) into Eq. (9.20) gives the normalized current on switches S_2 and S_3:

$$\overline{I_{S23\,RMS}} = \frac{I_{S23\,RMS}}{I'_o} = \sqrt{\frac{6 - 5\,\overline{I'_o}}{12}} \tag{9.21}$$

The curve of the normalized current $\overline{I_{S14\,RMS}}$ is shown in Fig. 9.13.

(C) Output Diodes Average Current

The output diodes average current is given by

$$I_{D1234} = \frac{1}{T_s} \int_0^{\Delta t_{32}} \left(I'_o - \frac{\overline{I'_o}}{\Delta t_{32}} t \right) dt \tag{9.22}$$

Thus, after integration in relation to time we find

$$I_{D1234} = I'_o \frac{(D - D_{ef})}{8}$$

(9.23)

Substituting Eqs. (9.12) in (9.23) it can be obtained

$$\overline{I_{D1234}} = \frac{I_{D1234}}{I'_o} = \frac{\overline{I'_o}}{8}$$

(9.24)

(D) Average Current in Clamping Diodes D_5 and D_6

The clamping diodes average current is

$$I_{Dc12} = \frac{1}{T_s} \int_0^{\Delta t_{21}} I'_o \, dt$$

(9.25)

Hence, after integration in relation to time and appropriate manipulation it can be found

$$\overline{I_{Dc12}} = \frac{I_{Dc12}}{I'_o} = \frac{(1 - D)}{2}$$

(9.26)

The corresponding curves are plotted in Fig. 9.14.

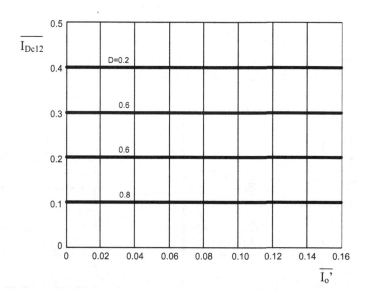

Fig. 9.14 Clamping diodes average current, as a function of the parameterized output current, taking D as a parameter

9.4 Commutation Analysis

In this section, commutation capacitors are added in parallel with the switches and dead time is considered, so the soft commutation phenomenon may be analyzed. Four commutation time intervals are added to one switching period. An appropriate design of the commutation capacitors can enable zero voltage switching (ZVS) for wide load range condition.

(A) Commutation of Switches S_1 and S_4

The first commutation time interval takes place between time intervals Δt_1 and Δt_2 described in Sect. 9.2. During time interval Δt_1 the capacitors C_1 and C_2 are discharged and capacitors C_3 and C_4 are both charged with initial voltage equal to $+V_i/2$. At the instant the switch S_1 is turned off, capacitor C_1 starts to charge with constant current and its voltage rises linearly towards $+V_i/2$. Simultaneously, capacitors C_3 and C_4 are discharged at constant current with their voltages decreasing linearly to $+V_i/4$. The converter topological state for this commutation time interval, along with the relevant waveforms are shown in Fig. 9.15.

During this commutation time interval, the voltages across commutation capacitors C_1, C_3 and C_4 are represented by Eqs. (9.27) and (9.28).

$$v_{C1}(t) = \frac{I'_o t}{1.5\,C} \tag{9.27}$$

$$v_{C3}(t) + v_{C4}(t) = V_i - \frac{I'_o t}{1.5\,C} \tag{9.28}$$

where $C = C_1 = C_2 = C_3 = C_4$.

A similar commutation takes place in the second half of the switching period, between time intervals Δt_5 and Δt_6 described in Sect. 9.2. This commutation time interval and its main waveforms are shown in Fig. 9.16.

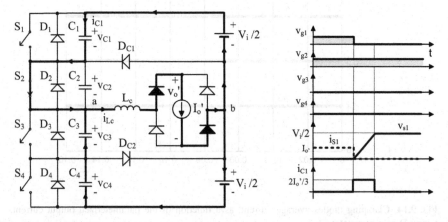

Fig. 9.15 Topological state and main waveforms for the first commutation time interval

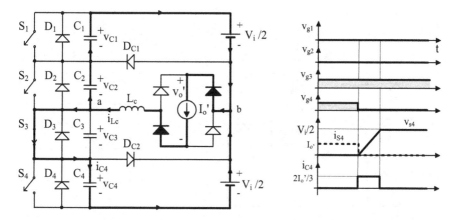

Fig. 9.16 Topological state and main waveforms for the third commutation time interval

(B) Commutation of Switches S_2 and S_3

The two more critical commutations take place when the inner switches (S_2 or S_3) are turned off.

First analyzing the commutation of switch S_2 that starts at the instant it is gated off. The converter topological state for this time interval is shown in Fig. 9.17. Before the beginning of this commutation, capacitor C_2 is discharged and capacitors C_3 and C_4 are both charged with voltages equal to $+V_i/4$.

At the instant the switch S_2 is turned off, capacitor C_2 starts to charge and its voltage rise towards $+V_i/2$ in a resonant way. Simultaneously, the capacitors C_3 and C_4 also discharge in a resonant way. If, by the end of the dead time, the voltage in capacitor C_2 does not reach $V_i/2$, the soft commutation is not achieved.

This commutation time interval and its main waveforms are presented in Fig. 9.17.

The voltage across capacitor C_2 evolves according to Eq. (9.29) during this time interval

$$v_{C2}(t) = \sqrt{\frac{L_r}{1.5\,C}}\,I'_o\,\text{sen}\,(\omega_o t) \tag{9.29}$$

where $\omega_o = \dfrac{1}{\sqrt{1.5 L_e\,C}}$.

In order to complete the charge of the capacitor C_2 and consequently achieve the desired commutation under zero voltage, the following constraint must be satisfied:

$$I'_o\sqrt{\frac{L_c}{1.5\,C}} \geq \frac{V_i}{2} \tag{9.30}$$

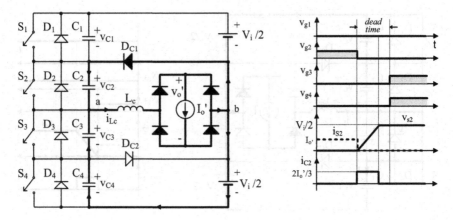

Fig. 9.17 Topological state and main waveforms for the second commutation time interval

Thus, the minimum necessary load current reflected to the transformer primary winding is

$$I'_{o\,crit} = \frac{V_i}{2} \sqrt{\frac{1.5\,C}{L_c}} \tag{9.31}$$

A similar commutation takes place in the second half of the switching period, between time interval Δt_6 and $\Delta t_{7,}$ described in Sect. 9.2. The time diagram, the relevant waveforms and topological state for the commutation time interval are shown in Fig. 9.18.

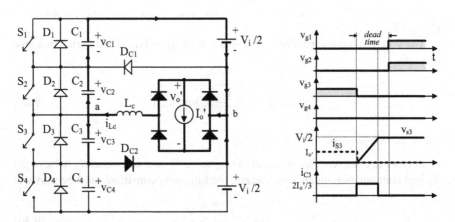

Fig. 9.18 Topological state and main waveforms for the fourth commutation time interval

The commutation process or the voltage transition across the switch ends after a time interval given by Eq. (9.31) that was deduced in the previous chapter, for the FB-ZVS-PWM dc-dc converter. The same equation is valid for the TL-ZVS-PWM converter, as long as we substitute equivalent capacitance $C_{eq} = 2C$ with $C_{eq} = \frac{3C}{2}$.

$$\Delta t = \left[\frac{\pi}{2} - \cos^{-1}\left(\frac{V_i}{2I'_{o\,crit}}\sqrt{\frac{3\,C}{2\,L_c}}\right)\right]\sqrt{\frac{3\,CL_c}{2}} \tag{9.32}$$

The dead-time t_d between the gate signals must be longer than Δt to prevent short circuit. Thus,

$$t_d \geq \Delta t. \tag{9.33}$$

9.5 Simplified Design Methodology and an Example of the Commutation Parameters

In this section it will be presented a simplified design methodology and an example of the converter commutation parameters, using the analysis results obtained in the previous sections. The specifications for this illustrative exercise are shown in Table 9.1.

Let us begin by selecting a static gain $q = 0.8$. Thus, the output DC voltage referred to the primary side of the transformer (V'_o) is:

$$V'_o = \frac{V_i}{2}\,q = 200 \times 0.8 = 160 \text{ V}$$

The transformer turns ratio (n) is:

$$n = \frac{V'_o}{V_o} = \frac{160}{50} = 3.2$$

Table 9.1 Design specifications

Input DC voltage (V_i)	400 V
Output DC voltage (V_o)	50 V
Output DC current (I_o)	10 A
Output power (P_o)	500 W
Switching frequency (f_s)	40 kHz

The nominal load current reflected to the transformer primary winding is:

$$I_o' = \frac{I_o}{n} = \frac{10}{3.2} = 3.125 \text{ A}$$

Let us consider a duty-cycle reduction $\overline{I_o'}$ of 10%. Hence, the inductance of the series inductor L_c may be determined as follows:

$$L_c = \frac{\overline{I_o'}(V_i/2)}{4 f_s I_o'} = \frac{0.1 \times 200}{4 \times 40 \times 10^3 \times 3.125} = 40 \times 10^{-6} \text{ H}$$

Let $C = 222$ pF (capacitances of the commutation capacitors associated in parallel with the switches). Thus, the minimum load current reflected to the transformer primary winding, necessary to achieve soft commutation of the inner switches S_2 and S_3 is:

$$I_{o\,crit}' = \frac{V_i}{2} \sqrt{\frac{1.5\,C}{L_c}} = 200 \times \sqrt{\frac{1.5 \times 222 \times 10^{-12}}{40 \times 10^{-6}}} = 0.577 \text{ A}$$

This current is 18.6% of the rated current, and it transfers 93 W to the load. This is the minimum load power necessary to achieve full realization of the ZVS commutation.

The nominal duty-cycle is given by

$$D_{nom} = \frac{V_o'}{V_i/2} + \overline{I_o'} = \frac{160}{200} + 0.1$$

Hence,

$$D_{nom} = 0.9$$

The effective duty-cycle is

$$D_{ef} = D_{nom} - \overline{I_o'} = 0.9 - 0.1 = 0.8$$

The dead-time is given by

$$t_d \geq \left[\frac{\pi}{2} - \cos^{-1}\left(\frac{V_i}{2\,I_{o\,crit}'} \sqrt{\frac{3\,C}{2\,L_c}} \right) \right] \sqrt{\frac{3\,CL_c}{2}}$$

Thus,

$$t_d \geq \left[\frac{\pi}{2} - \cos^{-1} \left(\frac{400}{2 \times 0.577} \sqrt{\frac{3 \times 222p}{2 \times 40\mu}} \right) \right] \sqrt{\frac{3 \times 222p \times 40\mu}{2}}$$

$$t_d \geq 181.3 \text{ ns}$$

9.6 Simulation Results

The design example of the Three-Level ZVS-PWM converter shown in Fig. 9.19 was simulated to validate the analysis presented in Sect. 9.5, with a dead time of 200 ns. Figure 9.20 presents the AC voltage v_{ab} and the output voltage v_o' and the inductor current. As it may be observed, the inductor current linear time intervals make the rectifier diodes be short circuited ($v_o' = 0$), decreasing the effective duty cycle.

Table 9.2 presents theoretical and simulated parameters component stresses, validating the mathematical analysis.

The commutation of switches S_1 and S_2, for the nominal power, are presented in Figs. 9.21 and 9.23, respectively. A detail of the switches S_1 and S_2 turn off are presented in Figs. 9.22 and 9.24. As it may be observed soft commutation is achieved for both switches, and the inner switches S_2 and S_3 have a more critical commutation due to the dead time.

A detail of the switch S_2 turn off at the critical power is presented in Fig. 9.25. As it may be observed ZVS commutation is on its limit.

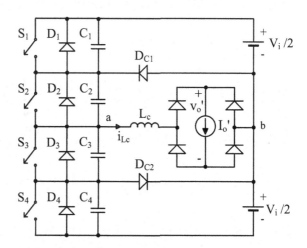

Fig. 9.19 Simulated converter

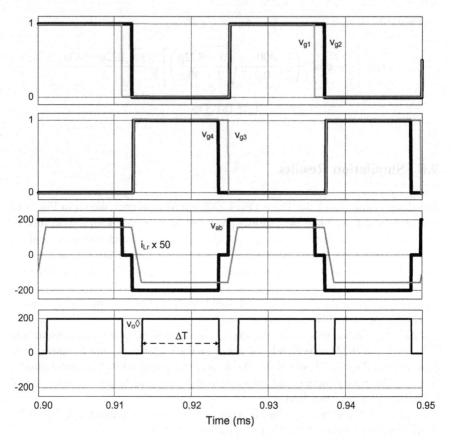

Fig. 9.20 Three-level ZVS-PWM converter simulation waveforms: switches drive signals, voltages v_{ab} and v'_o, and current i_{Lc}, for nominal power

Table 9.2 Theoretical and simulated results		Theoretical	Simulated
	V'_o (V)	160	159.9
	$i_{S14\ RMS}$ (A)	1.99	1.99
	$i_{S23\ RMS}$ (A)	2.12	2.12
	I_{D1234} (A)	0.04	0.036
	P_o (W)	500	499.9

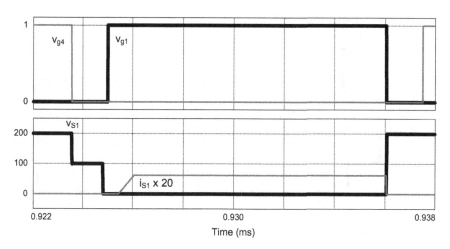

Fig. 9.21 Switch S_1 soft commutation at nominal power: S_1 and S_4 drive signals, voltage and current at switch S_1

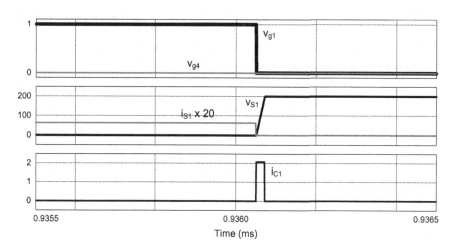

Fig. 9.22 A detail of switch S_1 turn off at nominal power: S_1 and S_4 drive signals, voltage and current at switch S_1 and capacitor C_1 current

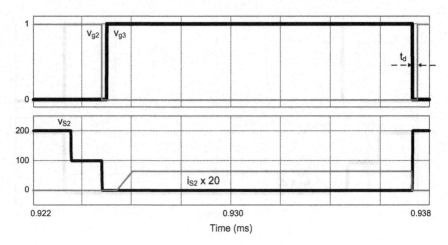

Fig. 9.23 Switch S_2 soft commutation at nominal power: S_2 and S_3 drive signals voltage and current at switch S_2

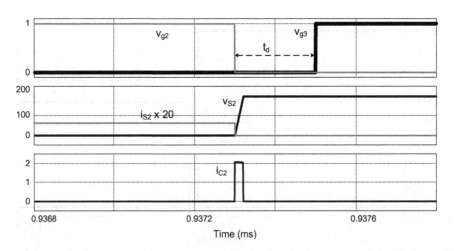

Fig. 9.24 A detail of switch S_2 turn off at nominal power: S_2 and S_3 drive signals, voltage and current at switch S_2 and capacitor C_2 current

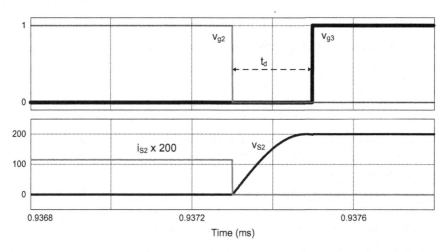

Fig. 9.25 A detail of switch S_2 turn off at the ZVS commutation limit: S_2 and S_3 drive signals, voltage and current at switch S_2

9.7 Problems

(1) The three-level NPC ZVS-PWM converter with an inductive output filter, along with its gate drive signals, is shown in Fig. 9.26. The converter has the following specifications:

$$V_i = 800 \text{ V} \quad N_p = 3N_s \quad f_s = 20 \times 10^3 \text{ Hz} \quad L_o = 200 \text{ } \mu\text{H} \quad C_o = 20 \text{ } \mu\text{F}$$
$$R_o = 2\Omega L_c = 120 \text{ } \mu\text{H} \quad \Delta T = 15 \text{ } \mu\text{s}$$

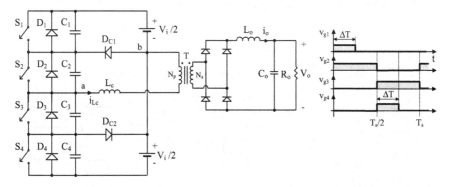

Fig. 9.26 Three-level NPC ZVS-PWM converter

Assuming that all components are ideal, calculate:

(a) The output voltage V_o, the output current I_o and the output power P_o;
(b) The output inductor current ripple;
(c) The duty-cycle loss.

Answers: (a) $V_o = 52$ V, $I_o = 26$ A, and $P_o = 1361$ W; (b) $\Delta I_{Lo} = 4$ A;
(c) $\Delta D = 0.21$

(2) The converter shown in Fig. 9.26 has commutation capacitors equal to $C = 4$ nF and large inductance of inductor L_o, so that its current ripple can be neglected. Calculate:

(a) The commutation time interval for switches S_1 and S_4;
(b) The commutation time interval for switches S_2 and S_3;
(c) The dead time to ensure soft commutation.

Answers: (a) $\Delta t_1 = 276$ nS; (b) $\Delta t_2 = 281$ ns; (c) $t_d \geq 281$ μs.

(3) For the same converter and parameters of exercise 2:

(a) Considering that ZVS commutation must be realized, determine the maximum value of resistance R_o and the output voltage and power on this condition;
(b) Simulate the converter and verify your results.

Answers: (a) $R_{o\ max} = 8.36\ \Omega$; $V_o = 71$ V; $P_o = 602$ W.

(4) The three-level NPC ZVS-PWM with a capacitive output filter, with its gate drive signals, is shown in Fig. 9.27. Consider all components ideal.

(a) Describe the converter operation, representing the topological states and the main waveforms;
(b) Obtain the static gain equation and compare with the static gain of the FB ZVS-PWM with a capacitive output filter studied in chapter VII;
(c) Simulate the converter and verify your results.

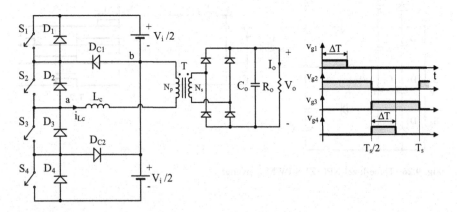

Fig. 9.27 Three-level NPC ZVS-PWM converter with output capacitive filter

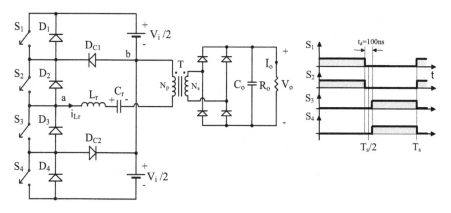

Fig. 9.28 Three-level NPC series resonant converter

(5) The converter shown in Fig. 9.27 has the following parameters:

$$V_i = 800 \text{ V} \qquad V_o = 200 \text{ V} \quad N_p = N_s$$
$$f_s = 50 \times 10^3 \text{ Hz} \quad L_c = 30 \text{ μH} \quad \Delta T = 8 \text{ μs}$$

Calculate the value of the load current I_o.
Answer: $I_o = 27.7$ A

(6) A Three-level NPC Series Resonant converter, with its gate drive signals and ideal components, is shown in Fig. 9.28. The converter operates with a constant switching frequency and has the following parameters:

$$V_i = 800 \text{ V} \quad N_p = N_s \quad C_r = 52.8 \text{ nF} \quad L_r = 12 \text{ μH} \quad R_o = 20 \text{ Ω} \quad C_o = 10 \text{ μF}$$

(a) Calculate the load voltage V_o, the output power P_o, the resonant capacitor peak voltage $V_{cr \, peak}$ and the peak value of the current in the inductor L_r;
(d) Describe the converter's operation, by representing the topological states and the main waveforms over a switching cycle.

Answers (a) $V_o = 400$ V, $P_o = 8$ kW, $V_{cr \, peak} = 473.61$ V, $I_{Lr \, peak} = 31.416$ A.

Reference

1. Pinheiro, J.R., Barbi, I.: The Three-Level ZVS-PWM DC-to-DC converter. IEEE Trans. Power Electron. **8**(4), 486–492 (1993)

Chapter 10
Asymmetric Half-Bridge ZVS-PWM Converter

Nomenclature

V_i	Input dc voltage
V_o	Output dc voltage
P_o	Output power
C_o	Output filter capacitor
L_o	Output filter inductor
R_o	Output load resistor
ZVS	Zero voltage switching
q	Static gain
D	Duty cycle
D_{nom}	Nominal duty cycle
f_s	Switching frequency
T_s	Switching period
t_d	Dead time
T	Transformer
n	Transformer turns ratio
v'_o (V'_o)	Output DC voltage referred to the transformer primary side, and its average value
C_{e1}, C_{e2}	DC bus capacitors
C_{eq}	Equivalent DC bus capacitors $(C_{eq} = C_{e1} + C_{e2})$
v_{Ce1}, v_{Ce2} (V_{Ce1}, V_{Ce1})	DC bus capacitors voltage and its average value
Δv_{Ceq}	Equivalent DC bus capacitors voltage ripple
i_o	Output current
I'_o	Average output current referred to the primary side of the transformer
I'_{ocrit}	Critical average output current referred to the primary side of the transformer
I_1	Magnetizing inductance current at the end of time interval 1
I_2	Magnetizing inductance current at the end of time interval 4
S_1 and S_2	Switches

© Springer International Publishing AG, part of Springer Nature 2019
I. Barbi and F. Pöttker, *Soft Commutation Isolated DC-DC Converters*,
Power Systems, https://doi.org/10.1007/978-3-319-96178-1_10

I_{S1} and I_{S2}	Switches commutation current, respectively
v_{g1} and v_{g2}	Switches S_1 and S_2 gate signals, respectively
D_1, D_2	Diodes in anti-parallel to the switches (MOSFET—intrinsic diodes)
C_1, C_2	Capacitors in parallel to the switches (MOSFET—intrinsic capacitors)
C	Capacitance in parallel to the switches $C = C_1 = C_2$
L_c	Commutation inductance
i_{Lc}	Commutation inductor current
v_{Lc}	Commutation inductor voltage
L_m	Transformer magnetizing inductance
i_{Lm}	Transformer magnetizing inductance current
I_{Lm}	Transformer magnetizing inductance average current
Δi_{Lm}	Transformer magnetizing inductance current ripple
v_{ab}	AC voltage, between points "a" and "b"
v_{S1}, v_{S2}	Voltage across the switches
i_{S1}, i_{S2}	Current on the switches
i_{C1}, i_{C2}	Current in the capacitors
Δt_a	Time interval of the first and second step of operation
Δt_b	Time interval of the fourth and fifth step of operation
Δt_c	Time interval of the third second step of operation
Δt_d	Time interval of the sixth step of operation
$I_{S1\ RMS}\ \left(\overline{I_{S1RMS}}\right)$	Switch S_1 RMS current and its normalized value
$I_{S2\ RMS}\ \left(\overline{I_{S2RMS}}\right)$	Switch S_2 RMS current and its normalized value

10.1 Introduction

The conventional half-bridge converter topology is widely used in middle power level applications, due to its simplicity. Just as the double-ended forward converter, it subjects the power switches to only the DC-bus input voltage and not twice that as the singled forward converter. Its major advantage over the double-ended forward converter is its symmetric operation, which means basically that the transformer average magnetizing current is zero.

However, despite these well-known advantages, the power switches commutate under hard-commutation, which implies in commutation losses and reduction of the converter efficiency.

To provide soft-commutation under ZVS in the half-bridge converter, the asymmetric (complementary) control was proposed to achieve ZVS operation for both switches [1]. Two drive signals are complementarily generated and applied to the power switches so that they may be turned on at ZVS conditions due to the fact

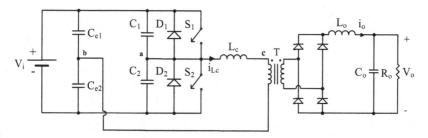

Fig. 10.1 Asymmetric half-bridge ZVS-PWM converter

that the transformer leakage inductance primary charges and discharges the commutation capacitances.

The asymmetric half bridge ZVS-PWM converter is shown in Fig. 10.1.

The DC bus capacitors C_{e1} and C_{e2} have different DC voltages due to the asymmetric drive signals. It will be shown later that the voltages across these capacitors are given as a function of the duty-cycle D, by Eqs. (10.1) and (10.2), respectively.

$$V_{Ce2} = D\,V_i \qquad (10.1)$$

$$V_{Ce1} = (1 - D)\,V_i \qquad (10.2)$$

The converter will be analyzed for $0 \le D \le 0.5$, because its behavior is the same for $0.5 \le D \le 1$.

10.2   Circuit Operation

In this section, it will be described the operation of the half-bridge converter shown in Fig. 10.2. To simplify, it will be assumed that:

- all components are ideal;
- the converter operates on steady-state condition;

Fig. 10.2 Asymmetric half-bridge ZVS-PWM converter

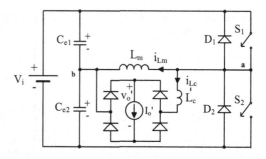

Fig. 10.3 Topological state
during time interval Δt_1

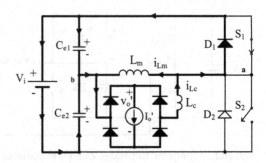

- the output filter is replaced by a DC current source I_o', whose value is the output average current referred to the primary winding of the transformer;
- The converter is controlled by asymmetric pulse width modulation (PWM);
- The magnetizing inductance is large enough so that its current ripple can be neglected.

The operation of the converter is described over a switching cycle, which is divided into six time intervals, each one representing a different topological state.

(A) Time Interval Δt_1 ($t_0 \leq t \leq t_1$)

At $t = t_0$, when switch S_2 is turned off, diode D_1 starts to conduct, as shown in Fig. 10.3. During this time interval energy is returned to the DC bus. The voltage v_{ab} is equal to $+V_{CE1}$ and the inductor current increases linearly until it reaches zero.

(B) Time Interval Δt_2 ($t_1 \leq t \leq t_2$)

This time interval, shown in Fig. 10.4, begins at the instant the commutation inductor current reaches zero. At this instant the switch S_1 starts to conduct and the inductor current i_{Lc} increases linearly until it reaches $+I_o'$ +iLm.

(C) Time Interval Δt_3 ($t_2 \leq t \leq t_3$)

The topological state for the time interval Δt_3 is shown in Fig. 10.5. This time interval starts at $t = t_2$, when the inductor current reaches $+I_o'$ +iLm. Switch S_1 conducts the current, the voltage v_{ab} is equal to V_{Ce1} and energy is transferred from the DC bus to the load. This time interval ends when switch S_1 is turned off and switch S_2 is turned on.

Fig. 10.4 Topological state
during time interval Δt_2

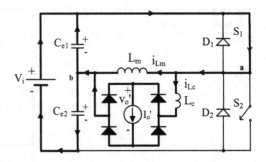

Fig. 10.5 Topological state
during time interval Δt_3

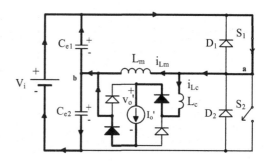

Fig. 10.6 Topological state
during time interval Δt_4

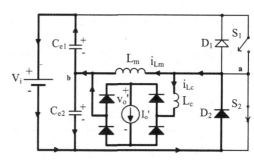

(D) Time Interval Δt_4 ($t_3 \leq t \leq t_4$)

At $t = t_3$ switch S_1 is turned off and diode D_2 starts to conduct the current, as shown
in Fig. 10.6. Once again, the commutation inductor delivers energy to the DC bus.
The voltage v_{ab} is equal to $-V_{Ce2}$ and the inductor current i_{Lc} decreases linearly and
reaches zero at the instant $t = t_4$.

(E) Time Interval Δt_5 ($t_4 \leq t \leq t_5$)

This time interval, shown in Fig. 10.7, begins at the instant t_4, when the commutation inductor current reaches zero. At this instant switch S_2 starts to conduct and
the inductor current decreases linearly.

(F) Time Interval Δt_6 ($t_5 \leq t \leq t_6$)

The topological state for the time interval Δt_6 is shown in Fig. 10.8. It starts at t_5
when the inductor current reaches $-I_0' + i_{Lm-}$. Switch S_2 conducts the current, the

Fig. 10.7 Topological state
during time interval Δt_5

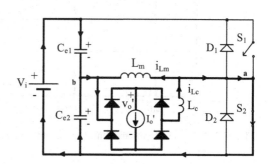

Fig. 10.8 Topological state during time interval Δt_6

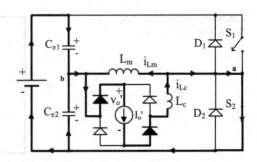

voltage v_{ab} is equal to $-V_{Ce2}$ and energy is transferred from the DC bus to the load. This time interval ends at t_6, when switch S_2 is turned off and switch S_1 is gated on.

The main waveforms along with the time diagram, over a switching cycle, are shown in Fig. 10.9.

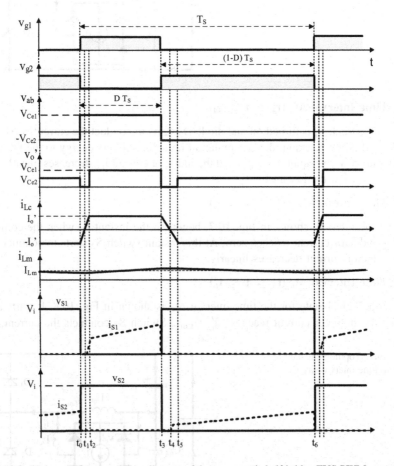

Fig. 10.9 Main waveforms and time diagram of the asymmetric half-bridge ZVS-PWM converter

10.3 Mathematical Analysis

(A) Output Characteristics

During the time intervals Δt_a and Δt_b, as shown in Fig. 10.10, the inductor current varies linearly, the voltage at the output of the rectifier stage is zero so no power is transferred to the load. As the voltage applied to the inductor is asymmetric, the rates of the inductor current i_{Lc} are unequal in the two time intervals.

During the time interval Δt_a the inductor L_c is subjected to the voltage given by Eq. (10.3).

$$v_{Lc} = (1 - D)\, V_i \tag{10.3}$$

The duration of this time interval is

$$\Delta t_a = \frac{2\, I_o'\, L_c}{(1 - D)\, V_i} \tag{10.4}$$

In a similar way, the voltage across the inductor L_c during the time interval Δt_b is given by

$$v_{Lc} = D\, V_i \tag{10.5}$$

and the duration of this time interval is

$$\Delta t_b = \frac{2\, I_o'\, L_c}{D\, V_i} \tag{10.6}$$

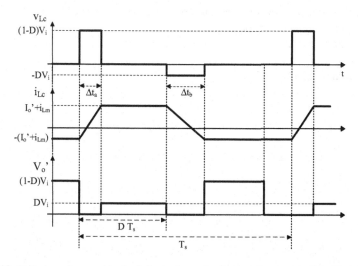

Fig. 10.10 Voltage and current in the inductor Lc and voltage at the output of the diode rectifier

The average value of the output voltage is determined by

$$V_o' = \frac{1}{T_s} \left[\int_{\Delta t_a}^{DT_s} (1-D) V_i \, dt + \int_{DT_s + \Delta t_b}^{T_s} D V_i \, dt \right] \qquad (10.7)$$

Integrating Eq. (10.7) yields

$$V_o' = \frac{1}{T_s} \{(1-D)V_i(D.T_s - \Delta t_a) + D V_i [(1-D) T_s - \Delta t_b]\} \qquad (10.8)$$

With appropriate algebraic manipulation, the DC voltage gain can be found by Eq. (10.9).

$$q = \frac{V_o'}{V_i} = \left[2D(1-D) - \frac{4 I_o' L_c f_s}{V_i} \right] \qquad (10.9)$$

The duty cycle loss, defined by Eq. (10.18), is proportional to the output current.

$$\overline{I_o'} = \frac{4 I_o' L_c f_s}{V_i} \qquad (10.10)$$

After substituting Eq. (10.10) into (10.9) we find the static gain given by

$$q = 2 D(1-D) - \overline{I_o'} \qquad (10.11)$$

The asymmetric half bridge ZVS-PWM converter output characteristics (static gain as a function of the normalized output current, taking the duty cycle as a parameter) are shown in Fig. 10.11. As it may be noticed, the average output voltage (V_o') depends on the load current (I_o') as occurred with converters already studied in the previous chapters.

The half bridge ZVS-PWM converter conversion ratio, considering $L_c = 0$, is shown in Fig. 10.12. As there are two duty-cycle values for a given static gain, it is usually limited to D = 0.5.

Fig. 10.11 Output characteristics of the half-bridge ZVS-PWM converter

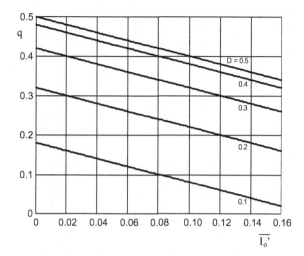

Fig. 10.12 Voltage transfer characteristics of the half-bridge ZVS-PWM converter

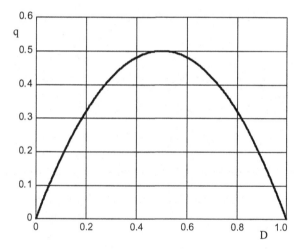

(B) **Magnetizing Inductance Average Current**

Due to the asymmetric operation, the magnetizing inductance average current is not zero and the transformer operates with magnetic flux bias, like occurs with all asymmetric isolated dc–dc converters.

Figure 10.14 shows the typical waveform of the magnetizing inductance current on steady-state condition for the asymmetrical half-bridge PWM converter presented in Fig. 10.13.

It was already demonstrated that the time intervals Δt_a and Δt_b are determined by Eqs. (10.12) and (10.13), respectively.

Fig. 10.13 Asymmetrical
half-bridge ZVS-PWM
converter

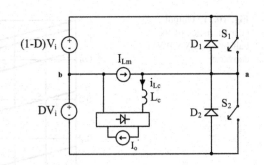

$$\Delta t_a = \frac{2 \, L_c \, I_o'}{(1 - D) V_i} \tag{10.12}$$

$$\Delta t_b = \frac{2 \, L_c \, I_o'}{D \, V_i} \tag{10.13}$$

Defining

$$\Delta t_c = D \, T_s - \Delta t_a = D \, T_s - \frac{2 \, L_c \, I_o'}{(1 - D) \, V_i} \tag{10.14}$$

and

$$\Delta t_d = (1 - D) T_s - \Delta t_b = (1 - D) \, T_s - \frac{2 \, L_c \, I_o'}{D \, V_i} \tag{10.15}$$

The average value of the current i_{Lc}, as shown in Fig. 10.14 is determined by

$$I_{Lc} = I_o' \, \frac{\Delta t_d - \Delta t_c}{T_s} \tag{10.16}$$

After substitution of Eqs. (10.14) and (10.15) in Eq. (10.16), and knowing that the average value of the magnetizing current I_{Lm} is equal to I_{Lc}, we find

$$I_{Lm} = I_o'(1 - 2 D) \left[1 - \frac{2 \, L_c \, f_s \, I_o'}{V_i \, D \, (1 - D)} \right] \tag{10.17}$$

As may be noticed in (10.17), the magnetizing inductance average current is zero only when the duty cycle is 0.5, condition in which the converter is on symmetrical operation. If the commutation inductance (L_c) is neglected, $I_{Lm} = I_o'(1 - 2 D)$.

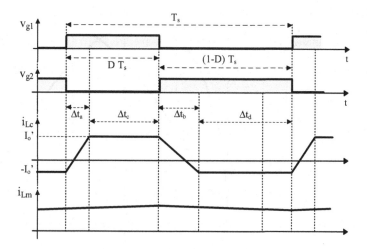

Fig. 10.14 Waveforms for analysis of the magnetizing current

(C) Switches RMS Current

According to the definition, the switches S_1 and S_2 RMS currents are determined by Eqs. (10.18) and (10.19) respectively.

$$I_{S1RMS} = \sqrt{\frac{1}{T_s} \int_0^{DT_s} \left[2\left(1-D\right)I_o'\right]^2 dt} \tag{10.18}$$

$$I_{S2RMS} = \sqrt{\frac{1}{T_s} \int_0^{(1-D)T_s} \left[2\,D\,I_o'\right]^2 dt} \tag{10.19}$$

After integration and appropriate algebraic manipulation, Eqs. (10.20) and (10.21) can be found

$$\overline{I_{S1RMS}} = \frac{I_{S1RMS}}{I_o'} = 2\left(1-D\right)\sqrt{D} \tag{10.20}$$

$$\overline{I_{S2RMS}} = \frac{I_{S2RMS}}{I_o'} = 2\,D\sqrt{1-D} \tag{10.21}$$

The switches S_1 and S_2 RMS currents, as a function of the duty cycle, are plotted in Fig. 10.15. The maximum RMS current for switch S_1 takes place when the duty cycle is 0.333 and for switch S_2 when the duty cycle is 0.667. If the converter operates with D = 0.5, the switches RMS currents become identical.

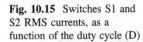

Fig. 10.15 Switches S1 and
S2 RMS currents, as a
function of the duty cycle (D)

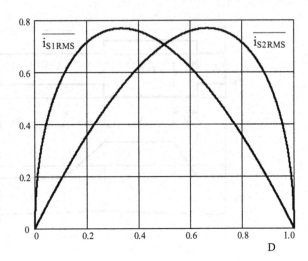

10.4 Commutation Analysis

In this section, the capacitors are added in parallel to the switches and a dead time is considered, so the soft commutation phenomenon may be analyzed. Four commutation time intervals are added to one switching period.

Due to the converter asymmetrical operation, the commutations take place under different conditions.

10.4.1 Switch S_1 Commutation

The first commutation starts at the instant switch S_1 is turned off and switch S_2 is not turned on yet (dead time). It has one linear time interval, followed by a resonant one, as shown in Fig. 10.16. In the linear time interval, the inductor current is I_o', and capacitors C_1 and C_2 are charged/discharged in a linear way, until capacitor C_1 reaches the voltage $(1 - D) V_i$ leading v_{ab} to zero. When v_{ab} reaches zero, the output rectifier is short circuited and the resonant time interval begins, with the inductor current and the capacitors voltage evolving in a resonant way, until the capacitor C_2 is completely discharged.

(A) Linear Time Interval (Δt_{1a})

The peak value of the magnetizing inductance can be mathematically expressed as

$$I_1 = I_{Lm} + \frac{\Delta i_{Lm}}{2} \tag{10.22}$$

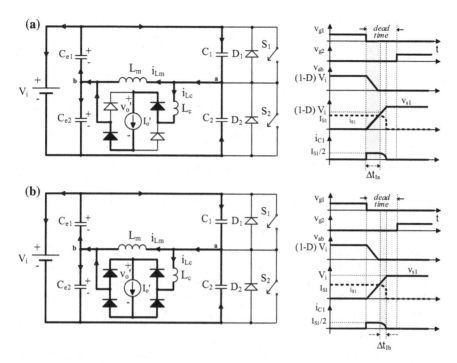

Fig. 10.16 Topological state and main waveforms for the switch S1 commutation: **a** linear time interval and **b** resonant time interval

where the value of Δi_{Lm} is given by

$$\Delta i_{Lm} = \frac{V_i}{2 L_m f_s} \times D(1 - D) \tag{10.23}$$

The average value of the magnetizing current is

$$I_{Lm} = I'_o \times (1 - 2D) \times \left(1 - \frac{2 L_c f_s I'_o}{V_i D (1 - D)}\right) \tag{10.24}$$

The current in switch S_1 at the instant of the commutation is mathematically represented by

$$I_{S1} = I'_o + I_1 \tag{10.25}$$

Substitution of Eqs. (10.22), (10.23) and (10.24) in Eq. (10.25) yields

$$I_{S1} = I'_o + I'_o \times (1 - 2D) \times \left(1 - \frac{2\,L_c\,f_s\,I'_o}{V_i\,D(1-D)}\right)$$

$$+ \frac{V_i}{2\,L_m\,f_s} \times D\,(1-D) \tag{10.26}$$

During the commutation linear time interval, the switch S_1 voltage evolves from zero to $(1 - D)\,V_i$. The duration of this time interval is

$$\Delta t_{1a} = \frac{C(1-D)V_i}{I_{C1}} \tag{10.27}$$

where I_{C1} is $I_{S1}/2$ and $C = C_1 = C_2$.

(E) Resonant Time Interval (Δt_{1b})

The equivalent circuit for the resonant time interval is presented in Fig. 10.17.
   The initial conditions are: $i_{Lc}(0) = I_{S1}$, $v_{C1}(0) = (1 - D)V_i$ and $v_{C2}(0) = DV_i$.
   The state plane trajectory for this resonant time interval is shown in Fig. 10.18.
   The radius of the semi-circle, the characteristic impedance and the angular frequency are given by Eqs. (10.28), (10.29) and (10.30), respectively.

$$R_1 = z\,I_{S1} \tag{10.28}$$

$$z = \sqrt{\frac{L_c}{2\,C}} \tag{10.29}$$

Fig. 10.17 Equivalent circuit for the resonant time interval

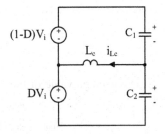

Fig. 10.18 State plane for the resonant time interval

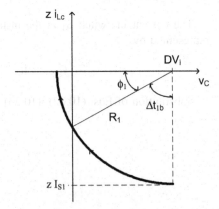

$$\omega = \frac{1}{\sqrt{2\,C\,L_c}} \tag{10.30}$$

Substitution of Eq. (10.26) in (10.28) gives

$$R_1 = z\left(I'_o + I_{Lm} + \frac{\Delta i_{Lm}}{2}\right) \tag{10.31}$$

By inspection, from Fig. 10.18 we can write

$$\omega\,\Delta t_{1b} + \phi_1 = \pi/2 \tag{10.32}$$

Thus

$$\phi_1 = \cos^{-1}\left(\frac{D\,V_i}{R_1}\right) \tag{10.33}$$

And

$$\Delta t_{1b} = \frac{1}{\omega}\left\{\frac{\pi}{2} - \cos^{-1}\left[\frac{D\,V_i}{z\left(I'_o + I_{Lm} + \Delta i_{Lm}/2\right)}\right]\right\} \tag{10.34}$$

The duration of the commutation is

$$\Delta t_1 = \Delta t_{1a} + \Delta t_{1b} \tag{10.35}$$

Hence, the dead time must satisfy the constraint

$$t_{d1} \geq \Delta t_{1a} + \Delta t_{1b} \tag{10.36}$$

10.4.2 Switch S_2 Commutation

In a similar way, the second commutation starts at the instant switch S_2 is turned off and switch S_1 is not turned on yet (dead time). It also has one linear time interval, followed by a resonant time interval, as shown in Fig. 10.19. In the linear time interval, the inductor current is I'_o, and capacitors C_1 and C_2 are discharged/charged in a linear way, until capacitor C_2 reaches the voltage DV_i leading v_{ab} to zero. When v_{ab} reaches zero, the output rectifier is short circuited and the resonant time interval begins, with the inductor current and the capacitors voltage evolving in a resonant way, until capacitor C_1 is completely discharged. At the resonant time interval, the magnetizing inductance current is subtracted from the load current, so the commutation process takes longer compared to the switch S_1 commutation.

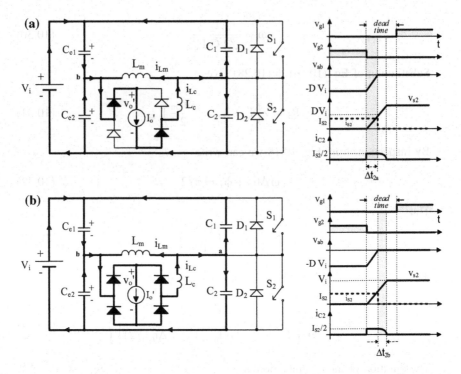

Fig. 10.19 Topological state and main waveforms for the switch S2 commutation: **a** linear time interval and **b** resonant time interval

(A) Linear Time Interval (Δt_{2a})

The magnetizing current at the instant of the commutation can be mathematically expressed as

$$I_2 = I_{Lm} - \frac{\Delta i_{Lm}}{2} \tag{10.37}$$

The switch current at the same instant is

$$I_{S2} = I_o' - I_2 = I_o' - \left(I_{Lm} - \frac{\Delta i_{Lm}}{2} \right) \tag{10.38}$$

Hence,

$$I_{S2} = I_o' - I_o' \times (1 - 2\,D) \times \left(1 - \frac{2\,L_c\,f_s I_o'}{V_i\,D(1-D)} \right)$$
$$+ \frac{V_i}{2\,L_m f_s} \times D(1-D) \tag{10.39}$$

During the commutation linear time interval, switch S_2 voltage evolves from zero to ($D V_i$). The duration of this time interval is

$$\Delta t_{2a} = \frac{C D V_i}{I_{C2}}$$

(10.40)

where I_{C2} is $I_{S2}/2$.

(B) Resonant Time Interval (Δt_{2b})

The equivalent circuit for the resonant time interval is presented in Fig. 10.20. The initial conditions are: $i_{Lc}(0) = I_{S2}$, $v_{C1}(0) = (1 - D)V_i$ and $v_{C2}(0) = DV_i$. The state plane trajectory for this resonant time interval is shown in Fig. 10.21. The radius of the semi-circle is given by Eq. (10.41).

$$R_2 = z I_{S2}$$

(10.41)

Thus,

$$R_2 = z\left[I'_o - \left(I_{Lm} + \frac{\Delta i_{Lm}}{2}\right)\right]$$

(10.42)

For the realization of the soft-commutation, the radius R_2 must satisfy the constraint

Fig. 10.20 Equivalent circuit for the resonant time interval

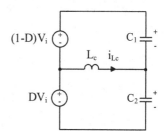

Fig. 10.21 State plane for the resonant time interval

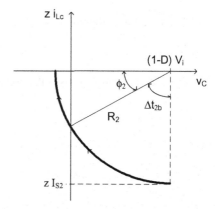

$$R_2 \geq (1 - D)V_i \tag{10.43}$$

Substituting Eq. (10.42) into (10.43) yields

$$z\left[I'_o - \left(I_{Lm} - \frac{\Delta i_{Lm}}{2}\right)\right] \geq (1 - D)V_i \tag{10.44}$$

As $I_{s2} < I_{s1}$ the switch S_2 commutation is the critical one. Appropriate manipulation of Eq. (10.44) gives

$$I'_{o\,min} = \frac{(1 - D)V_i}{z} + \left(I_{Lm} - \frac{\Delta i_{Lm}}{2}\right) \tag{10.45}$$

where $I'_{o\,min}$ is the minimum load current referred to the transformer primary winding, that ensures switch S_2 soft commutation.

By inspection of the graphic shown in Fig. 10.21 we can write

$$\omega\,\Delta t_{2b} + \phi_2 = \pi/2 \tag{10.46}$$

Thus

$$\phi_2 = \cos^{-1}\left[\frac{(1 - D)V_i}{R_2}\right] \tag{10.47}$$

Substitution of (10.47) into (10.46) gives

$$\omega\,\Delta t_{2b} = \frac{\pi}{2} - \cos^{-1}\left[\frac{(1 - D)V_i}{R_2}\right] \tag{10.48}$$

Hence,

$$\Delta t_{2b} = \frac{1}{\omega}\left\{\frac{\pi}{2} - \cos^{-1}\left[\frac{(1 - D)\,V_i}{I'_o - \left(I_{Lm} + \frac{\Delta i_{Lm}}{2}\right)}\right]\right\} \tag{10.49}$$

The duration of the commutation is

$$\Delta t_2 = \Delta t_{2a} + \Delta t_{2b} \tag{10.50}$$

The dead-time must satisfy the constraint

$$t_{d2} \geq \Delta t_2 \tag{10.51}$$

10.5 Simplified Design Methodology and an Example of the Commutation Parameters

In this section a design methodology and an example are presented according to the mathematical analysis presented in the previous sections. The converter is designed according to the specifications shown in Table 10.1.

Considering a transformer turns ratio (n) of 3.2, the output current (I'_o) and the output voltage (V'_o), both referred to the primary side of the transformer are calculated as follows.

$$I'_o = \frac{I_o}{n} = \frac{10}{3.2} = 3.125 \text{ A}$$
$$V'_o = V_i\, n = 50 \times 3.2 = 160 \text{ V}$$

The static gain is

$$q = \frac{V'_o}{V_i} = \frac{160}{400} = 0.4 \text{ V}$$

Considering a duty cycle reduction of 5%, the commutation inductor is given by.

$$L_c = \frac{\overline{I'_o}\, V_i}{4\, f_s\, I'_o} = \frac{0.05 \times 400}{4 \times 40 \times 10^3 \times 3.125} = 40\ \mu\text{H}$$

The nominal duty cycle is calculated using Eq. (10.19).

$$D_{nom} = \frac{1}{2} - \sqrt{\frac{V_i - 8\, I'_o\, L_c\, f_s - 2\, V_i\, q}{4\, V_i}} = 0.342$$

The DC Bus capacitors average voltages are:

$$V_{Ce1} = V_i(1 - D_{nom}) = 263.25 \text{ V}$$
$$V_{Ce2} = V_i D_{nom} = 136.75 \text{ V}$$

The magnetizing inductance average and ripple current are determined by Eqs. (10.17) and (10.23), respectively

Table 10.1 Design specifications

Input DC voltage (V_i)	400 V
Output DC voltage (V_o)	50 V
Output DC current (I_o)	10 A
Output power (P_o)	500 W
Switching frequency (f_s)	40 kHz

$$I_{Lm} = (1 - 2 D_{nom})I_o' \left(1 - \frac{2L_c f_s I_o'}{V_i D_{nom}(1 - D_{nom})}\right) = 0.878 \text{ A}$$

$$\Delta i_{Lm} = \frac{V_i}{L_m f_s} \times D_{nom}(1 - D_{nom}) = 0.562 \text{ A}$$

Considering the commutation capacitance $C = C_1 = C_2 = 4$ pF, switch S_1 commutation current and commutation time interval are calculated as follows:

$$\omega = \frac{1}{\sqrt{2 C L_c}} = 5.59 \times 10^6 \text{ rad/s}$$

$$z = \sqrt{\frac{L_c}{2 C}} = 223.6$$

$$I_{S1} = 2 I_{C1} = 2 \times \frac{(I_o' + I_{Lm} + \Delta i_{Lm}/2)}{2} = 4.285 \text{ A}$$

$$\Delta t_{1a} = \frac{C(1 - D_{nom}) V_i}{I_{C1}} = 49.15 \text{ ns}$$

$$\Delta t_{1b} = \frac{1}{\omega}\left\{\frac{\pi}{2} - \cos^{-1}\left[\frac{D V_i}{z(I_o' + I_{Lm} + \Delta i_{Lm}/2)}\right]\right\} = 25.62 \text{ ns}$$

$$\Delta t_1 = \Delta t_{1a} + \Delta t_{1b} = 74.77 \text{ ns}$$

Switch S_2 commutation current and commutation time interval are:

$$I_{S2} = 2 I_{C2} = 2 \times \frac{(I_o' - I_{Lm} + \Delta i_{Lm}/2)}{2} = 2.528 \text{ A}$$

$$\Delta t_{2a} = \frac{C(1 - D_{nom}) V_i}{I_{C2}} = 43.28 \text{ ns}$$

$$\Delta t_{2b} = \frac{1}{\omega}\left\{\frac{\pi}{2} - \cos^{-1}\left[\frac{(1 - D_{nom})V_i}{z(I_o' - I_{Lm} + \Delta i_{Lm}/2)}\right]\right\} = 86.66 \text{ ns}$$

$$\Delta t_2 = \Delta t_{2a} + \Delta t_{2b} = 129.9 \text{ ns}$$

The minimum dead time is equal to Δt_2.

$$t_{dmin} = \Delta t_2 = 129.9 \text{ ns}$$

10.6   Simulation Results

The design example of the asymmetrical half bridge ZVS-PWM converter, shown in Fig. 10.22, is simulated to validate the analysis presented in Sect. 10.5, with a dead time of 150 ns. Figure 10.23 presents the main waveforms for the nominal power. As it may be noticed, capacitor C_{e1} and C_{e2} average voltages are different, due to the asymmetry.

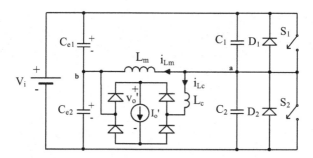

Fig. 10.22 Simulated converter

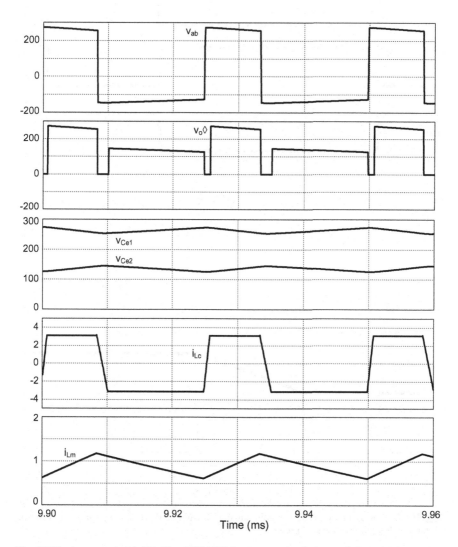

Fig. 10.23 Asymmetrical half bridge ZVS-PWM converter simulation waveforms: voltages vab, V'$_o$, V$_{Ce1}$ and V$_{Ce2}$ and currents iLc and iLm, for nominal power

Table 10.2 Theoretical and
simulated results

	Theoretical	Simulated
V_o' [V]	160	161.6
$i_{S1\ RMS}$ [A]	2.40	2.26
$i_{S2\ RMS}$ [A]	1.73	1.75
V_{Ce1} [V]	263	263.2
V_{Ce2} [V]	137	136.8

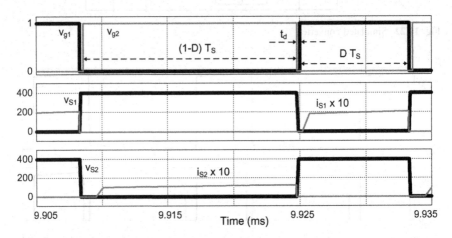

Fig. 10.24 Switches S1 and S2 commutation: S1 and S2 drive signals, voltage and current at the switches

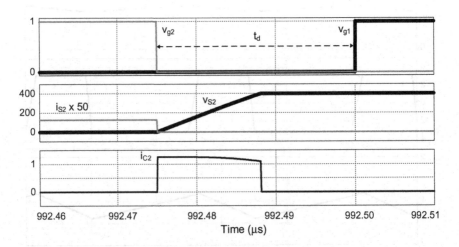

Fig. 10.25 A detail of switch S2 soft commutation: S1 and S2 drive signals, voltage and current at the switch and capacitor C2 current

Table 10.2 presents the theoretical and simulated parameters and the component stresses, validating the mathematical analysis.

The commutation of switch S_1 and S_2 are presented in Fig. 10.24. The current available for the switch S_2 commutation is smaller, so this is the critical commutation. A detail of the switch S_2 turn off is presented in Fig. 10.25. As it may be observed soft commutation is achieved.

10.7 Problems

(1) The asymmetrical ZVS-PWM converter with an inductive output filter is presented in Fig. 10.26. The converter has the following specifications:

$$V_i = 400 \text{ V} \quad N_p/N_s = 50 \text{ V} \quad R_o = 2 \ \Omega \quad f_s = 40 \times 10^3 \text{ Hz} \quad L_o = 200 \ \mu\text{H}$$
$$C_o = 50 \ \mu\text{F} \quad D = 0.3 \quad L_c = 10 \ \mu\text{H} \quad L_m = 2 \text{ mH}$$

Calculate:

(a) The average output voltage V_o, the output average output current I_o and the output power P_o.
(b) The average magnetizing inductance current.
(c) The average voltage in the DC Bus capacitors.
(d) The magnetizing inductance ripple current.

Answers: (a) $V_o = 31.1$ V, $I_o = 15.56$ A, $P_o = 484$ W; (b) $I_{Lm} = 1.15$ A; (c) $V_{Ce1} = 230$ V, $V_{Ce2} = 120$ V; (d) $\Delta i_{Lm} = 1.05$ A.

(2) Considering the asymmetrical ZVS-PWM converter presented in Fig. 10.27, with the commutation capacitors (C_1 and C_2) of 2 nF and a high leakage and magnetizing inductance, calculate the commutation time of both switches, with the following specifications:

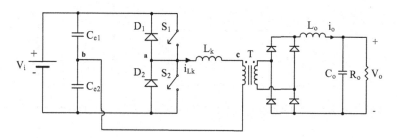

Fig. 10.26 Asymmetrical HB ZVS-PWM converter

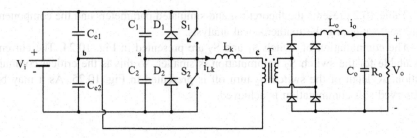

Fig. 10.27 Asymmetrical HB ZVS-PWM converter

$$V_i = 400 \text{ V} \quad N_p/N_s = 50 \text{ V} \quad R_o = 2 \, \Omega \quad f_s = 40 \times 10^3 \text{ Hz}$$
$$C_o = 50 \, \mu F \quad D = 0.3 \quad L_k = 10 \, \mu H$$

Answers: $t_{c1} = 252.5$ ns, $t_{c2} = 357$ ns.

(3) The asymmetrical ZVS-PWM converter presented in Fig. 10.27, has the following parameters:

$$V_i = 400 \text{ V} \quad N_p = N_s \quad C_1 = C_2 = 0.47 \text{ nF} \quad R_o = 2 \, \Omega \quad L_m = 10 \text{ mH}$$
$$f_s = 40 \times 10^3 \text{ Hz} \quad C_{e1} = C_{e2} = 10 \, \mu F \quad D = 0.3 \quad L_r = 20 \, \mu H$$

Calculate:

(a) The minimum output current ($I_{o \, min}$) to achieve ZVS commutation on both switches.
(b) The dead time to obtain ZVS commutation.

Answers: (a) $I_{o \, min} = 3.1$ A; (b) $t_d = 216$ ns.

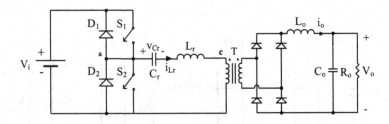

Fig. 10.28 Topological variation of the asymmetrical HB ZVS-PWM converter

(4) A topological variation of the asymmetrical ZVS-PWM is presented in Fig. 10.28. The converter has the following specifications:

$V_i = 400$ V $N_p = N_s = 1$ $R_o = 10\ \Omega$ $f_s = 40 \times 10^3$ Hz $L_o = 500\ \mu H$
$C_o = 50\ \mu F$ $D = 0.3$ $L_r = 20\ \mu H$ $Cr = 10\ \mu F$ $L_m = 20$ mH

Calculate:

(a) The resonant capacitor voltage ripple.
(b) The resonant capacitor average voltage.

Answers: (a) $\Delta V_{Cr} = 13.64$ V; (b) $V_{Cr} = 120$ V.

Reference

1. Imbertson, P., Mohan, N.: Asymmetrical duty cycle permits zero switching loss in PWM circuits with no conduction loss penalty. In: IEEE, pp. 1061–1066 (1991 October)

Chapter 11
Active Clamp ZVS-PWM Forward Converter

Nomenclature

V_i	Input DC voltage
V_o	Output DC voltage
P_o	Output power
C_o	Output capacitor filter
L_o	Output inductor filter
R_o	Output load resistor
ZVS	Zero voltage switching
q	Converter static gain
D	Duty cycle
D_{nom}	Nominal duty cycle
f_s	Switching frequency
T_s	Switching period
t_d	Dead time
n	Transformer turns ratio
T	Transformer
N_1, N_2 and N_3	Transformer windings
L_c	Commutation inductance (may be the transformer leakage inductance or an additional inductor, if necessary)
i_{Lc}	Commutation inductor current
L_m	Transformer magnetizing inductance
i_{Lm} (I_{Lm})	Transformer magnetizing inductance current and its average value
v_o' (V_o')	Output voltage referred to the transformer primary side and its average value
v_{Lm}	Magnetizing inductance voltage
V_{C3} ($\overline{V_{C3}}$)	Clamping capacitor voltage and its normalized value
I_o' ($\overline{I_o'}$)	Average output current referred to the primary and its normalized value
i_i (I_i)	DC bus current and its average value
i_{Lm} (I_{Lm})	Magnetizing inductor current and its average value
Δi_{Lm}	Magnetizing inductor current ripple

© Springer International Publishing AG, part of Springer Nature 2019
I. Barbi and F. Pöttker, *Soft Commutation Isolated DC-DC Converters*,
Power Systems, https://doi.org/10.1007/978-3-319-96178-1_11

I_1	Magnetizing inductor current at the end of time interval 4
I_2	Magnetizing inductor current at the end of time interval 8
S_1	Main switch
S_2	Active clamp switch
D_1, D_2	Diodes in anti-parallel to the switches (MOSFET—intrinsic diodes)
D_3, D_4	Output rectifier diodes
C_1, C_2	Capacitors in parallel to the switches (MOSFET—intrinsic capacitors)
v_{ab} (V_{ab})	Converter ac voltage, between points "a" and "b" and its average value
v_{S1}, v_{S2}	Voltage across switches
i_{S1}, i_{S2}	Switches current
i_{C1}, i_{C2}, i_{C3}	Capacitors current
Δt_1	Time interval of the first and second step of operation (t_2-t_0)
Δt_2	Time interval of the third and fourth step of operation (t_4-t_2)
Δt_3	Time interval of the fifth, sixth and seventh step of operation (t_7-t_4)
Δt_{32}	Time interval of the third step of operation in CCM (t_3-t_2)
$I_{S1\,RMS}$, $\left(\overline{I_{S1\,RMS}}\right)$	Switch S_1 RMS current and its normalized value

11.1   Introduction

The conventional forward converter is shown in Fig. 11.1. The isolating transformer T has an auxiliary winding (N_3) to provide the transformer demagnetization. However, the energy stored in the transformer leakage inductance (L_c) is not removed by the auxiliary winding. To prevent overvoltage across the switch S_1 at the instant it is turned off, a passive dissipative clamping circuit (RCD) is usually used to clamp the voltage and also to dissipate in the resistor R the energy stored in the leakage inductances. However, it is obvious that this solution sacrifices the efficiency of the converter.

Fig. 11.1 Power stage diagram of the forward converter with passive dissipative clamping circuit

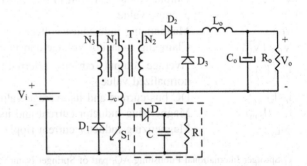

Fig. 11.2 Active clamp
forward converter

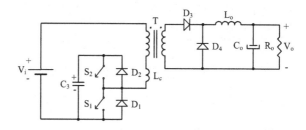

Fig. 11.3 Active clamp
ZVS-PWM forward converter

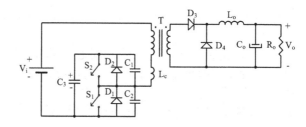

Figure 11.2 shows an active clamp forward converter [1]. Switch S_2 and capacitor C_3 are added to the original circuit to provide the transformer demagnetization and also to recycle the energy accumulated in the transformer leakage inductance to the input voltage source V_i. The switch operates at constant switching frequency with appropriate dead-time between the gate signals. The converter is pulse-width-modulated (PWM). The efficiency of the active clamp forward converter is higher than the conventional counterpart for the mentioned reasons.

If capacitors with appropriate capacitances are added in parallel to the switches S_1 and S_2, as shown in Fig. 11.3, ZVS commutation may be achieved [2], contributing even more to improve the converter's efficiency. The transformer leakage inductance is utilized in the soft commutation process and usually external inductors are not necessary.

11.2   Circuit Operation

In this section, the converter shown in Fig. 11.4, without the capacitors in parallel to the switches, is analyzed (the commutation analysis is presented in Sect. 11.4). The following assumptions are made to simplify:

- all components are ideal;
- the converter operates on steady-state;
- the output filter is replaced by a DC current source I_o', whose value is the output average current referred to the primary side of the transformer;
- The converter is controlled by pulse width modulation (PWM) without dead time between the two switches gate pulses.

Fig. 11.4 Active clamp
PWM Forward converter
referred to the transformer
primary side

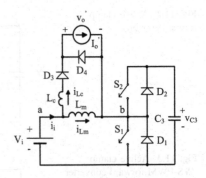

The operation of the converter is described during one switching period that is divided into eight time intervals, each one representing a different topological state of the converter.

(A) Time Interval Δt_1 ($t_0 \leq t \leq t_1$)

The time interval Δt_1, shown in Fig. 11.5, starts at $t = t_0$, when switch S_1 is gated on and switch S_2 is turned off. Although switch S_1 is gated on, current does not flow through it because the DC bus current i_1 is negative. Diode D_1 conducts the current and L_m is demagnetized while L_c is magnetized.

Fig. 11.5 Topological state
during time interval Δt_1

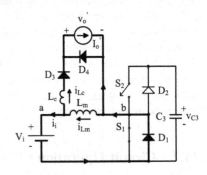

Fig. 11.6 Topological state
during time interval Δt_2

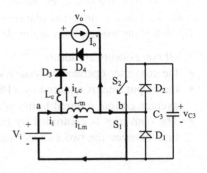

Fig. 11.7 Topological state
during time interval Δt_3

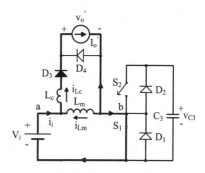

Fig. 11.8 Topological state
during time interval Δt_4

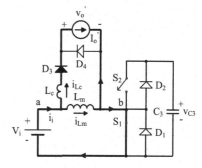

(B) Time Interval Δt_2 ($t_1 \leq t \leq t_2$)

This time interval, presented in Fig. 11.6, starts at $t = t_1$, when the current in the DC
bus (i_i) reaches zero and S_1 is conducting. In this time interval L_m continues
demagnetizing and L_c magnetizing.

(C) Time Interval Δt_3 ($t_2 \leq t \leq t_3$)

The time interval Δt_3, shown in Fig. 11.7, begins when the commutation inductance
current reaches I_o', turning off diode D_4. During this time interval L_m current
decreases linearly and the commutation inductor current is constant.

(D) Time Interval Δt_4 ($t_3 \leq t \leq t_4$)

This time interval, shown in Fig. 11.8, starts at $t = t_3$, when the magnetizing
inductance current reaches zero and increases linearly.

(E) Time Interval Δt_5 ($t_4 \leq t \leq t_5$)

The time interval Δt_5, shown in Fig. 11.9, starts at $t = t_4$, when switch S_2 is gated
on and switch S_1 is turned off. Although switch S_2 is gated on, current does not flow
through it because the DC bus current i_i is positive. Hence, diode D_2 conducts the
current, beginning another time interval, in which L_m and L_c are demagnetized by
linear currents. Diode D_4 shorts circuit the load.

Fig. 11.9 Topological state
during time interval Δt_5

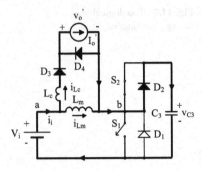

Fig. 11.10 Topological state
during time interval Δt_6

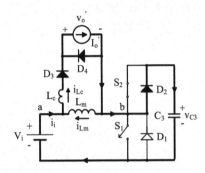

Fig. 11.11 Topological state
during time interval Δt_7

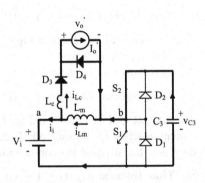

Fig. 11.12 Topological state
during time interval Δt_8

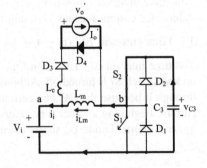

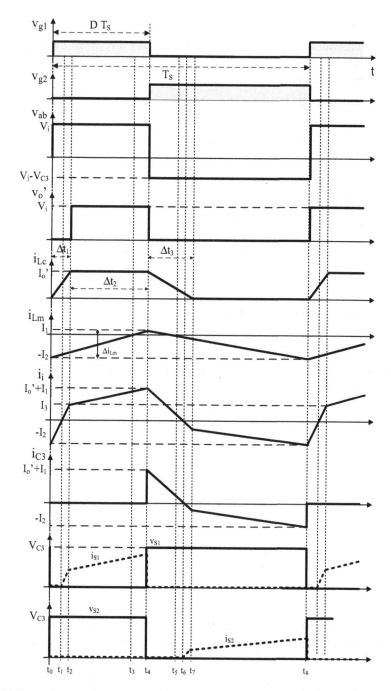

Fig. 11.13 Main waveforms and time diagram of the active clamp forward converter

(F) Time Interval Δt_6 ($t_5 \leq t \leq t_6$)

The time interval Δt_6, shown in Fig. 11.10, starts at $t = t_5$, when the magnetizing inductor current becomes negative. The commutation inductor current increases linearly during this time interval.

(G) Time Interval Δt_7 ($t_6 \leq t \leq t_7$)

This time interval, presented in Fig. 11.11, stars at $t = t_6$, when the current in the DC bus (i_i) reaches zero and S_2 is conducting. Currents through L_m and L_c evolve linearly.

(H) Time Interval Δt_8 ($t_7 \leq t \leq t_8$)

The time interval Δt_8, shown in Fig. 11.12, begins when the commutation inductance current reaches zero, turning off diode D_3. During this time interval L_m current increases linearly.

The main waveforms indicating the time intervals are shown in Fig. 11.13.

11.3   Mathematical Analysis

(A) Capacitor C_3 Average Voltage

As it can be observed in Fig. 11.13, the average value of the voltage across the magnetizing inductance is $V_{ab} = 0$. This voltage is given by

$$V_{ab} = \frac{1}{T_s}\left[\int_0^{DT_s} V_i\,dt + \int_0^{(1-D)T_s} (V_i - V_{C3})\,dt\right] = 0 \tag{11.1}$$

thus,

$$V_i DT_s = -(V_i - V_{C3})(1 - D)T_s \tag{11.2}$$

and

$$\overline{V_{C3}} = \frac{V_{C3}}{V_i} = \frac{1}{1 - D} \tag{11.3}$$

where $\overline{V_{C3}}$ is the normalized voltage across the clamping capacitor C_3.

Figure 11.14 shows $\overline{V_{C3}}$ as a function of the duty cycle (D).

Fig. 11.14 Normalized capacitor C_3 average voltage as a function of the duty cycle

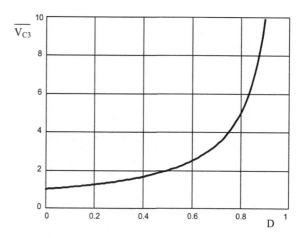

(B) Output Characteristics

The average output voltage V'_o is given by

$$V'_o = \frac{1}{T_s} \int_0^{\Delta t_2} V_i \, dt = \frac{1}{T_s} V_i \, \Delta t_2 \qquad (11.4)$$

The time intervals Δt_1 and Δt_2, shown in Fig. 11.13, are given by Eqs. (11.5) and (11.6) respectively.

$$\Delta t_2 = D \, T_s - \Delta t_1 \qquad (11.5)$$

$$\Delta t_1 = \frac{I'_o L_c}{V_i} \qquad (11.6)$$

Substituting (11.5) into (11.4) gives

$$q = \frac{V'_o}{V_i} = D - \overline{I'_o} \qquad (11.7)$$

where $\overline{I'_o} = \frac{I'_o L_c f_s}{V_i}$ is the duty-cycle loss.

As it may be noticed in Eq. (11.7), the duty cycle loss $\overline{I'_o}$ is proportional to the average output current I'_o. Figure 11.15 shows the output characteristics or the normalized output voltage as a function of the normalized output current taking the duty cycle as a parameter. Due to the duty cycle loss, the load voltage is dependent on the load current.

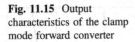

Fig. 11.15 Output characteristics of the clamp mode forward converter

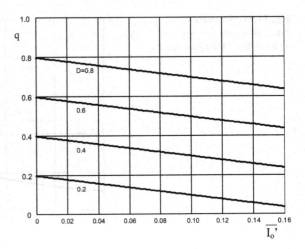

(C) **Switch S_1 RMS Current**

Switch S_1 RMS current is calculated disregarding the magnetizing inductance current and is given by

$$I_{S1\ RMS} = \sqrt{\frac{1}{T_s}\int_0^{\Delta t_2} I_o'^2\,dt} \qquad (11.8)$$

Hence,

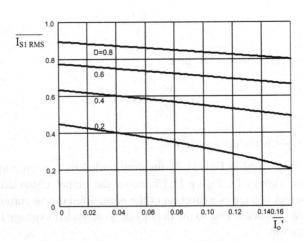

Fig. 11.16 Switch S_1 RMS normalized current, as a function of the normalized output current, taking the duty cycle as a parameter

$$\overline{I_{S1\,RMS}} = \frac{I_{S1\,RMS}}{I'_o} = \sqrt{D - I'_o} \qquad (11.9)$$

The switch S_1 RMS normalized current $\overline{I_{S1\,RMS}}$, as a function of the normalized output current, taking the duty cycle as a parameter, is shown in Fig. 11.16.

(D) Magnetizing Inductance Average Current

The forward converter operates asymmetrically. Consequently, the average magnetizing inductance current is not zero, as it may be noticed in the waveforms shown in Fig. 11.17.

Applying the Kirchhoff current law to the average current of the circuit shown in Fig. 11.4 it can be found

$$I_1 = I_{Lc} + I_{Lm} \qquad (11.10)$$

where I_1, I_{Lc}, and I_{Lm} are the average values of the input, commutation and magnetizing inductance currents.

The average commutation inductor current is obtained by inspection from Fig. 11.17 and is given by

$$I_{Lc} = I'_o\,D + I'_o\,\frac{\Delta t_3}{2T_S} - I'_o\,\frac{\Delta t_1}{2T_S} \qquad (11.11)$$

From the commutation inductor voltage equation, (11.12) may be written to obtain Δt_3:

$$\Delta t_3 = \frac{(1-D)}{DV_i}\,L_c I'_o \qquad (11.12)$$

According to the energy conservation principle we have

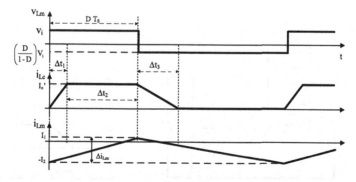

Fig. 11.17 Waveforms of: magnetizing inductor voltage (v_{Lm}), commutation inductor current (i_{Lc}), and magnetizing inductor current (i_{Lm})

$$I_i V_i = I_o' V_o' \tag{11.13}$$

Substituting (11.7) into (11.13) gives

$$I_i = I_o' \left(D - \frac{f_s L_c I_o'}{V_i} \right) \tag{11.14}$$

Substituting (11.6), (11.12) and (11.14) into (11.10) yields

$$I_{Lm} = -\frac{f_s L_c I_o'^2}{2DV_i} \tag{11.15}$$

which represents the average value of the transformer magnetizing current.

11.4   Commutation Analysis

In order to analyze the commutation process of the power switches, the commutation capacitors are added in parallel with the switches and a dead time between the power switches gate signal is considered. Four commutation time intervals are found in a switching period.

Due to the converter asymmetrical operation, the commutations take place under different conditions. At the instant switch S_1 is turned off, the current available to charge/discharge the commutation capacitors is the load current plus the magnetizing inductance current (I_1), as shown in Fig. 11.18. However, at the instant switch S_2 is turned off, the current available to charge/discharge the commutation capacitors is only the magnetizing inductance current (I_2), since the load is short circuited by diode D_4, as shown in Fig. 11.19. Hence, the second commutation is more critical than the first one. Both commutations take place under zero voltage

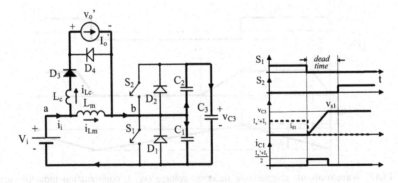

Fig. 11.18 Topological state and main waveforms for the first commutation time interval

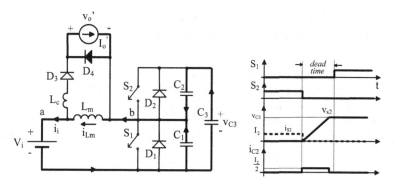

Fig. 11.19 Topological state and main waveforms for the second commutation time interval

(ZVS), providing the appropriate parameter combination is use in the converter design stage.

The magnetizing inductance current ripple is given by (11.16).

$$\Delta i_{Lm} = V_i \frac{D}{(1-D)} \frac{T_s(1-D)}{L_m} = \frac{V_i D T_s}{L_m} \tag{11.16}$$

The commutation current I_2 is calculated as follows:

$$I_2 = I_{Lm} + \frac{\Delta i_{Lm}}{2} \tag{11.17}$$

Hence

$$I_2 = \frac{f_s L_c I_o'^2}{2 V_i} + \frac{V_i D}{2 L_m f_s} \tag{11.18}$$

The minimum dead time needed to ensure ZVS commutation of switch S_2 (the critical commutation) is determined by

$$t_d = \frac{C V_{C3}}{I_2/2} \tag{11.19}$$

where $C_1 = C_2 = C$.

11.5   Simplified Design Methodology and an Example of the Commutation Parameters

In this section a design methodology and a simplified design example are provided, using the mathematical analysis results obtained in the previous sections. The converter specifications are shown in Table 11.1.

In this simplified design example, it was chosen the transformer turns ration = 3.2. The output current $\left(I'_o\right)$ and the output voltage $\left(V'_o\right)$, both referred to the primary side of the transformer, are

$$I'_o = \frac{I_o}{n} = \frac{10}{3.2} = 3.125 \text{ A}$$

$$V'_o = V_i \, n = 50 \times 3.2 = 160 \text{ V}$$

The static gain is given by

$$q = \frac{V'_o}{V_i} = \frac{160}{400} = 0.4 \text{ V}$$

Consider a duty cycle loss of 5%. The commutation inductance, consequently, is given by

$$L_c = \frac{\overline{I'_o} V_i}{f_s I'_o} = \frac{0.05 \times 400}{40 \times 10^3 \times 3.125} = 160 \text{ μH}$$

According to (11.7), the duty cycle at the operation point is given by

$$D = q + \overline{I'_o}$$

Hence,

$$D = 0.45$$

Table 11.1 Design specifications

Input DC voltage (V_i)	400 V
Output DC voltage (V_o)	50 V
Output DC current (I_o)	10 A
Output power (P_o)	500 W
Switching frequency (f_s)	40 kHz

The average voltage of capacitor C_3 is calculated as follows

$$V_{C3} = \frac{V_i}{1 - D_{nom}} = \frac{400}{1 - 0.45} \cong 728 \text{ V}$$

Switch S_1 RMS current is calculated

$$I_{S1 \text{ RMS}} = I'_o \sqrt{D - \overline{I'_o}} = 3.125 \times \sqrt{0.45 - 0.05} = 1.97 \text{ A}$$

Assume that the magnetizing inductance of the transformer is 4 mH. Hence, the magnetizing inductance current ripple is:

$$\frac{\Delta i_{Lm}}{2} = \frac{V_i D}{2 L_m f_s} = \frac{400 \times 0.45}{2 \times 4 \times 10^{-3} \times 40 \times 10^3} = 0.56 \text{ A}$$

The average value of the magnetizing inductance current is

$$I_{Lm} = \frac{-f_s L_c I_o'^2 V_i D}{2 D V_i} = \frac{-40 \times 10^3 \times 160 \times 10^{-6} \times 3.125^2}{2 \times 0.45 \times 400} = -0.1736 \text{ A}$$

The current available for the commutation of the switch S_2 is:

$$|I_2| = I_{Lm} + \frac{\Delta i_{Lm}}{2} = 0.1736 + 0.56 = 0.7336 \text{ A}$$

Assume that the capacitances of the commutation capacitors are $C_1 = C_2 = 200$ pF. Consequently, the minimum dead time needed to accomplish ZVS commutation is

$$t_d = \frac{2CV_{C3}}{I_2} = \frac{2 \times 200 \times 10^{-12} \times 728}{0.7336} = 0.397 \text{ }\mu s$$

11.6 Simulation Results

The active clamp ZVS-PWM forward converter, shown in Fig. 11.20, is simulated at the specified operation point to verify the analysis and the calculation presented in the design example of Sect. 11.5, with a dead time of 450 ns. Figure 11.21 shows the voltage v_{ab} and the output voltage v'_o, capacitor C_3 voltage, the commutation inductor current and the magnetizing inductance current.

Table 11.2 compares theoretical and simulated results, validating the mathematical analysis.

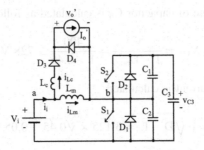

Fig. 11.20 Simulated converter

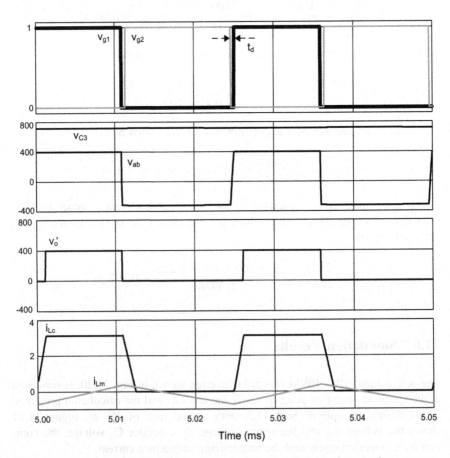

Fig. 11.21 Active clamp ZVS-PWM forward converter simulation waveforms: voltages v_{ab}, v_{C3} and v'_o, commutation inductance current i_{Lc} and magnetizing inductance current i_{Lm}

Table 11.2 Theoretical and simulated results

	Theoretical	Simulated
V'_o (V)	160	160.16
$i_{S1\ RMS}$ (A)	1.97	1.91
V_{C3} (V)	728	719.8
I_2 (A)	0.73	0.79

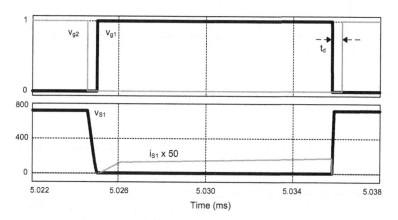

Fig. 11.22 Switch S_1 voltage and current waveforms

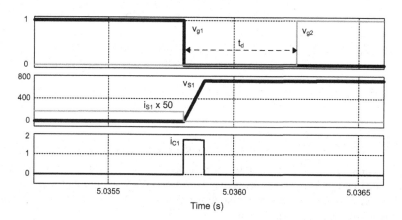

Fig. 11.23 A detail of the switch S_1 commutation: S_1 and S_2 drive signals, voltage and current at switch S_1 and capacitor C_1 current

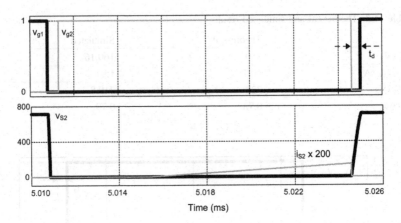

Fig. 11.24 Switch S_2 voltage and current waveforms

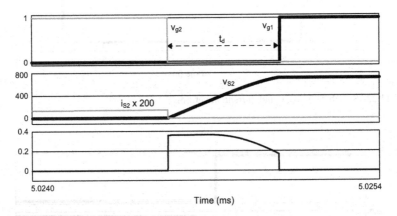

Fig. 11.25 A detail of the switch S_2 commutation: S_1 and S_2 drive signals, voltage and current at switch S_2 and capacitor C_2 current

The commutation at switches S_1 and S_2 are presented in Fig. 11.22 and in Fig. 11.24, respectively. A detail of the switches S_1 and S_2 turn off are presented in Fig. 11.23 and in Fig. 11.25. As it may be observed soft commutation is achieved on both switches. The current available for the switch S_2 commutation is smaller, so this is the critical commutation.

11.7 Problems

(1) The active clamp forward converter shown in Fig. 11.26 has the following specifications:

$$V_i = 100\,V \quad C_3 = 10\,\mu F \quad R_o = 2\,\Omega \quad f_s = 50 \times 10^3\,Hz \quad L_m = 1\,mH$$
$$C_o = 10\,\mu F \quad D = 0.4 \qquad L_r = 10\,\mu H \quad L_o = 500\,\mu H \qquad N_p = N_s$$

(a) Describe the topological states and plot the relevant waveforms for one switching period.
(b) Calculate the capacitor C_3 average voltage.
(c) Calculate V_o' and I_o' and the output power.
(d) Calculate the magnetizing inductance average current.
(e) Calculate the magnetizing inductance current ripple.

Answers: (b) $V_{C3} = 166.7\,V$; (c) $V_o' = 32\,V$, $I_o' = 16\,A$, $P_o = 512\,W$; (d) $I_{Lm} = 1.6\,A$; (e) $\Delta i_{Lm} = 4\,A$.

(2) Considering the converter presented in Fig. 11.26 with commutation capacitors of 2 nF in parallel to the switches and disregarding the inductance L_o current ripple, calculate:

(a) The switches current at the commutation instant.
(b) The commutation time intervals.

Answers: (a) $i_{S1} = 15.7\,A$, $i_{S2} = 354\,A$; (b) $\Delta_{t1} = 42.46\,ns$, $\Delta_{t2} = 188\,ns$.

(3) Figure 11.27 shows a topological variation of the active clamp forward converter.

(a) Describe the topological states and plot the main waveforms for one switching period.
(b) Obtain the equation for the capacitor C_3 average voltage.
(c) Find the voltage across the power switches assuming that $\Delta V_{C3} = 0$.

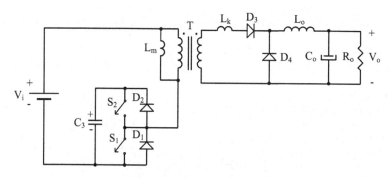

Fig. 11.26 Active clamp forward converter

Fig. 11.27 Active clamp
forward converter topological
variation

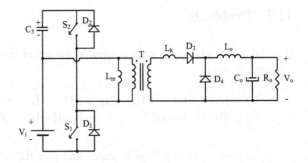

(d) Find the output voltage average value for $N_p = N_s$.
(e) Discuss the advantages and disadvantages of this topology compared to the
topology shown in Fig. 11.3.
(f) Simulate the proposed converter.

Answers: (b) $V_{C3} = \frac{D}{(1-D)} V_i$; (c) $V_S = \frac{V_i}{(1-D)}$; (d) $V_o = DV_1 - \frac{L_k f_s I_o}{V_i}$

(4) The asymmetrical ZVS-PWM converter shown in Fig. 11.27 has the following
parameters:

$$V_i = 400 \text{ V} \quad N_p = N_s \quad C_1 = C_2 = 0.47 \text{ nF} \quad R_o = 2 \text{ } \Omega \quad L_m = 10 \text{ mH}$$
$$f_s = 40 \times 10^3 \text{ Hz} \quad C_{e1} = C_{e2} = 10 \text{ } \mu\text{F} \quad D = 0.3 \quad L_r = 20 \text{ } \mu\text{H}$$

Calculate:

(a) The minimum output current ($I_{o \text{ min}}$) to achieve ZVS commutation on both
switches.
(b) The required dead time to obtain ZVS commutation.

Answers: (a) $I_{o \text{ min}} = 3.1 \text{ A}$; (b) $t_d = 216 \text{ ns}$.

(5) In the active clamp forward converter shown in Fig. 11.28, the leakage inductor
is represented in the primary winding of the transformer.

(a) Describe the topological states and plot the relevant waveforms for one
switching period.
(b) Assuming that $N_p = N_s$, find the equation for the static gain and the
equation for the average voltage across the clamping capacitor C_3.

Answers: (b) $V_o = DV_1 - \frac{L_r f_s I_o}{V_i}$; $V_{C3} = \frac{V_i}{(1-D)}$

Fig. 11.28 Active clamp
forward converter with the
leakage inductance in the
primary side of the
transformer

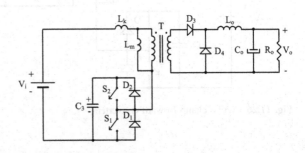

Fig. 11.29 Active clamp
flyback converter

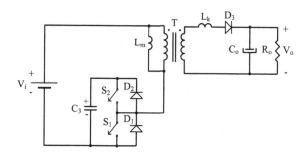

(6) An active clamp flyback converter [3] is shown in Fig. 11.29.

   (a) Describe the topological states and plot the main waveforms for one
       switching period.
   (b) Find the output voltage equation, considering $N_p = N_s$.
   (c) Find the average current in the transformer magnetizing inductance.

Answers: (b) $V_o = V_i \dfrac{D - \Delta D}{[1 - (D - \Delta D)]}$, where $\Delta D = \dfrac{2L_r f_s I_o}{V_i}$; (c) $I_{Lm} = \dfrac{I_o}{1 - (D - \Delta D)}$

References

1. Jitaru, I.D., Cocina, G.: High efficiency DC-DC converter. In: IEEE APEC 1994, pp. 638–644
   (1994)
2. Duarte, C.M.C., Barbi, I.: A family of ZVS-PWM active-clamping DC-to-DC converters:
   synthesis, analysis, design, and experimentation. IEEE Trans. Circ. Syst. **44**(8) (1997)
3. Watson, R., Lee, F.C., Hua, G.C.: Utilization of an active-clamp circuit to achieve soft
   switching in flyback converters. IEEE Trans. Power Electron., 162–169 (1996)

Fig. 11.29 Active clamp
flyback converter

(b) An active clamp flyback converter [3] is shown in Fig. 11.29.

References

Printed in the United States
By Bookmasters